W9-BNT-345

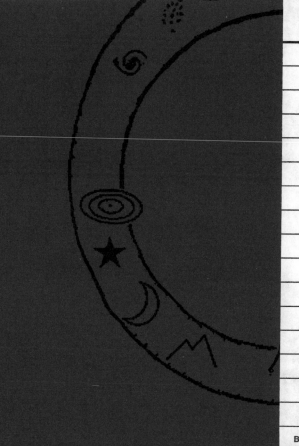

DATE DUE

APR 18 2011	

BRODART, CO.   Cat. No. 23-221

In Praise of Science

In Praise of Science

Curiosity, Understanding, and Progress

Sander Bais

The MIT Press   Cambridge, Massachusetts   London, England

First published in the Netherlands by Amsterdam University Press, Amsterdam

Printed in Italy

Design and layout: Gijs Mathijs Ontwerpers, Amsterdam

ISBN 978-0-262-01435-9

Library of Congress Cataloging-in-Publication Data
Bais, Sander.
 In praise of science : curiosity, understanding, and progress / Sander Bais.
    p. cm.
 Includes index.
 ISBN 978-0-262-01435-9 (hardcover : alk. paper)
 1. Science. I. Title.
 Q158.5.B35 2010
 500—dc22
                        2009035675

Contents

Preface 7

This book is written too late, because I only wrote it *after* having given many courses on physics and science. But it was the teaching itself that inspired me to actually do it. The book may serve as a starting point for a broad freshman type course on the Turning points in Science which emphasizes the unity and the connectivity of the natural sciences. I hope that from this particular perspective, the book will bring science closer to a larger audience and will contribute to the battle against scientific illiteracy. Besides a bird's eye view of the sciences, some chapters are concerned with how science comes about, how it works, and why it is such a central ingredient of human culture.

I'd like to thank many friends and colleagues for often long and inspiring discussions. In the first place my colleagues together with whom I have given and developed interdisciplinary courses like *Turning points* and *Big Questions* at the University and the University College of Amsterdam, also those of the *European Comenius Course* on Natural Science in Cambridge: John Barrow, Roel van Driel, Michiel van der Klis, Frank Linde, Steph Menken, Simon Conway Morris. I thank colleagues and visitors at the Santa Fe Institute for Complexity, for sharing their views varying from traffic jams, via emergence to econophysics and linguistics: Doyne Farmer, David Krakauer, Eric Smith, Harold Morowitz, Murray Gell-Mann, George Soros, Cormack Mc Carthy, Jeffrey West and others. Further, I thank my colleagues of the Institute for Theoretical Physics in Amsterdam: Jan Smit, Robbert Dijkgraaf, Kareljan Schoutens, Erik Verlinde, and Jan Pieter van der Schaar for discussions. I learned from discussions with Frits van Oostrom, Michiel Stokhuijzen, Sander Mooij, Vera de Vries and my children.

Without the active support of the teams of Amsterdam University Press and MIT Press this book would not have come into being, therefore I want to thank, Erik van Aert, Jaap Wagenaar, Arnout van Omme, Martin Voigt, Gita Devi Manaktala and Susan Clark. I am indebted to Anne Löhnberg for an intense and efficient transatlantic collaboration on the multilingual preparation of the manuscript.

Santa Fe,
October 2009

Our soul can only transcend the ordinary when it has traveled along the boundaries of the entire cosmos and has looked from above upon our small assortment of pieces of land, mostly immersed in water, which are largely uninhabitable even where they rise above the surface, because of the extreme heat or cold. It was said: "Is this the place that had to be divided by fire and sword among so many peoples? Oh, how ridiculous the divisions we mortals make!"

Seneca

Introduction

This book talks about science as a human enterprise. We talk about curiosity, or how science comes about, about understanding, or what the fundamental insights are that science has brought us, and about progress, or the way science is matched by technology and how science is perceived by the public. So besides describing what science has achieved and might achieve, the book touches on the poignant discrepancy in our cultural life between science and established culture. Science is about decoding discernable reality – in other words: Nature – and as there is just one Nature, in the end there can only be one science. Science has led to a vital chain of irreversible steps in our collective consciousness and as such determines, to a larger extent than is usually assumed, what human culture is. I do not refer as much to the huge socioeconomic impact of technological advances, as to the cultural dimension and meaning of science per se. Perhaps more than anything else, fundamental science has shaped human perceptions of the world that both surrounds us and is inside us. It has given us a clear account of what our place in the cosmos is and how it came about. Indeed one could go so far as to claim that science comes close to defining what it means to be a human being. Once we realize how profound this impact is, we can be puzzled and even worried about the ever-increasing marginalization of the natural sciences in the public arena, in other words the problem of scientific illiteracy. By ignoring the conceptual content of the sciences, we run the risk as individuals of becoming uncritical citizens. As for the collective, this ignorance will severely limit our contributions in the struggle to create a better world. A better world not based on the quest for power or money, but rather on the quest for truth and the ethics derived from it.

Indeed, science has managed to be expelled to the outskirts of our daily lives. Scientists form a closed society: they spend their lives in exotic and possibly scary laboratory environments, or in ivory towers where many people feel like *personæ non gratæ*.

If we include the technological dimension in our discussion of the impact of science on society, the picture becomes even more striking. Modern technological developments have been so overwhelming that some people think the roles have been reversed: reality is now trying to imitate imagination, rather than the other way around. Maybe we should be concerned about maintaining the appropriate level of technical skills to ensure a strong innovative potential in an evermore competitive world. A world, that in spite of an increasing overdose of cultural conflicts, appears inescapably driven to become a single global entity.

The process of globalization does not necessarily imply that diversity has to disappear. After all, in the global village a pizza will not spontaneously turn into a Peking duck. Fortunately, people enjoy partaking in a rich variety of cultural delights, by trying the most exotic local cuisines, sampling very diverse schools of thought, or just by traveling to other continents. The real threat of globalization lies in cultural intolerance. In order to attain a peaceful coexistence in the cultural domain, there is a need for us as individuals and as a society to develop an overall frame of reference. A need for a contemporary culture which should be the expression of an ever-increasing body of fundamental, evidence-based knowledge. Science offers a helping hand by saying: Don't turn away from undeniable facts, and don't commit crimes against logic. One indication of the role science can play in a global culture is the fact that fundamental scientific knowledge does not depend on ethnic, political or religious background. Quite the opposite: science is a process through which curiosity liberates us from the chains of collective myths and prejudices. It has shown the ability to transcend historical limitations. That's why I maintain hopeful: knowledge leads to awareness which in turn leads to reasonable actions.

Transcending the mythical narrative does not mean eradicating myths, but rather reducing them to their proper historical and factual proportions. After all, these myths – whether they are based on ethnicity, gender, religion or nationality – constitute an undeniable component of reality. They act as a collective memory and can be a source of information and inspiration that should not be ignored. It should rather be understood from a proper historical and evolutionary perspective, which may or may not shrink their societal impact to more modest proportions. The challenge is not to forget, but rather to come to terms with our past. Meanwhile, we should ask ourselves whether such mythical baggage should ever be allowed to prevail over the laws of society, or whether it belongs instead in the public domain, subjected to the public debate just like other expressions of human culture such as art and science. Dearest Sleeping Beauty, here is a prince who would love to kiss you …

What I have just put forward translates into a perspective on education. In many of the world's high-school systems teenagers have to choose the subjects or profiles that attract them most: art and culture, science and technology, or something in between: economy and society. Many choose the latter. For most children this choice comes too early. We tend to order things on a straight line from art to science, from creativity to rigorous logic, or, in a more orthodox view, from female to male. All this bears witness to a rather suffocating narrow-mindedness. It suggests that these subjects are opposites, that science is the opposite of culture; one might get the impression that they are mutually exclusive. Culturally minded people don't like science,

and scientifically minded people are insensitive to art. Culture freaks hate mathematics, because they find it too hard, while scientists are mostly nerds, indifferent to the subtleties of the human soul because they have a hard time dealing with their own emotions.

How strange! Perhaps we can start by placing all these endeavors of the human mind on a circle instead of on a line, or on the corners of a polygon. That would already drastically change the overall image of our civilization, and it would certainly lead to a more balanced partitioning where each of the educational profiles can be seen as an intermediate as well as an extreme. Maybe more importantly, it would open up numerous new possibilities, new interdisciplinary territories to explore, and new lessons to be learned and to be taught. This approach might liberate the minds of our children and enable them to discover and follow their true talents, making the pursuit of happiness less of a burden. It could even make the pursuit of education coalesce with the pursuit of happiness.

Allowing our pupils to discover what their abilities are appears to me to be one of the primary tasks of education. Often I say to my

11

students: "Your talents are the most valuable comrades in life, and the good thing is that they will stay with you." We all know that life can take strange and unforeseen turns, where you may lose what you thought was yours: you may lose your money, your possessions, and your partner along the way, as well as other persons dear to you, because they may move away or die. But somehow your talents will stay with you even in the most adverse circumstances. At such moments you may realize more than ever how precious and how indestructible talents are, providing you with the strength to survive and allowing you to rebuild yourself from what little is left. Consider the mental courage of people who have been imprisoned for very long times because of their political ideas or ethnic origins, like Primo Levi, Nelson Mandela, Joseph Brodsky or Aung San Suu Kyi. It is under dramatic circumstances that the true constitution of our personality becomes manifest, but it is that same constitution which governs our actions in ordinary life. So, my slogan is: *Just as your talents stay faithful to you, you should stay faithful to them.*

So far, so good. But is it really possible to bridge the gap between artistic and scientific talents in a single person through inspiring forms of education, or is this just one of those fashionable *fata morganas* overly optimistic educators want us to believe in? I believe it is possible. In fact, I think talent itself is what allows us to bridge gaps that initially look extremely wide and deep.

This book is divided into three parts. In the first chapter, I reflect on the roots of knowledge, and I point out how wonder and curiosity may liberate us from the iron embracement of prejudice, and this naturally leads to a demystification of our world view. Viewed sociologically, demystification can be a painful process, because at first it brings uncertainty and alienation. Short-term fears and conservative reactions to science should be contrasted with a more distant and long-term view, where we find demystification leading naturally to liberation and emancipation. Hopefully this book will convince you that the turning points in science – where our collective perspective had to take difficult and unexpected turns – are milestones in human evolution that have affected the emancipation of civilians, of workers, of women, and nowadays of children as well.

In the second part – Chapters II and III – of the book we descend into science itself. In Chapter II we focus on what drives science and scientists, and what makes science such a unique enterprise, exploiting virtually all human capabilities in a unique collective effort to understand the world outside and inside us. I point out some of the methodological aspects underlying the success of this amazingly robust knowledge-generating system. In Chapter III, I slowly build up a compact top-down view on the great turning points in our thinking about nature. I do so by exploiting an "Ouroboros of nature," a special, kaleidoscopic perspective, which facilitates a multi-layered survey of different aspects of these turning points in the sciences, while underscoring their mutual connectedness. Looking at the whole, one realizes that there is in fact only one science, and that these turning points stand for the historical instances where science could no longer go straight ahead, as it appeared to be stuck in a dead-end street. It is here that the miraculous mixture of creativity and ingenuity of the great scientists found the key to open a hidden door. Observations and abstract notions derived from these turn-ing points shook the very foundations of our vision of nature, forcing us too to take a fundamental turn in our thinking. At the same time these turns provide powerful examples of how a crisis can evoke a wave of creative thinking.

Scientific creativity is deeply rooted in the necessity to come to terms with irrefutable facts. Such struggles take place at the birth of any new paradigm, and only much later do we appreciate the profoundness of the consequences stemming from the disclosure of new facts which at first sight appeared rather innocuous. When I, as a curious young man, learned about these turning points, each time it seemed as if a veil was lifted from my eyes, bringing me closer to reality; a veil that would never return. Such is the irreversibility of growing insight.

That brings me to the third part of the book, where we return to the hard-to-overestimate impact that these turning points have had on the human condition and the human state of mind. Whereas in the second part I offer a straight exposé of the fundamental content of the natural sciences, in the third part (Chapter IV) I present a more personal view on the meaning and societal impact of science. While this chapter offers a critical analysis of how science is perceived in the public domain and how knowledge percolates down towards applications, at the same time it is a personal reflection on my life as a scientist. The increasing impact of science is contrasted with the fading image of the sciences in the public domain. Of course we are flooded with high-tech tools and gadgets which make life more interesting and enjoyable, but I would like to emphasize the dark side of this consumerism driven by ruthless economic principles. It allows people to slide into a jumble of ignorance, non-awareness and sometimes tragic isolation.

How come we do not recognize the great wealth of science as a source of insight and inspiration? Is that because applied technology is the only thing we see? Do we make mistakes in our judgments as a consequence of our ignorance? Is it our upbringing and the tacit transfer of values and prejudices? Or is a lousy schooling system to blame, which leaves our children clueless? Is it because we let them flounder on the Internet in a vast ocean of futilities? Of course the Internet itself is a splendid outcome of science, a phenomenal medium, but it is also a Nowhere Land where some guidance is necessary to not get lost. A Nowhere Land where virtuality turns into reality at an incredible rate. Is this a new means by which the tentacles of established culture reach even deeper into our personal lives, or is it a means for liberating a more universal spirit of our times? Does the Internet open our eyes, or is it blinding us, or maybe both? It makes the need for a discriminating mind even more acute.

Highly sophisticated manipulation by commerce and advertising has transformed a considerable fraction of the population into addicts with artificial needs. Democracy in practice gets stretched to the extreme because armies of well-paid lobbyists take over our capitals. This leaves science as a lonely orphan in the educational landscape: for nerds only. Whatever the answer to all those questions may be, I will argue that we should do something about it and that we can. The remaining question is which price we are willing to pay, in a mental as well as a material sense. Basically, the question is: What is the future worth to us?

Nauthilus, als Plinius seghet, es. i. wonder dat in de ze leghet. ii. langhe armen hevet voren, tusschen dien. ii., als wijd horen, es i vel dinne ende breet. So heffet hi hoghe up ghereet sine arme voren metten velle, so seilti henen als die snelle. Metten voeten roertet onder, metten starte stiertet, dats wonder. Comt hem vaer in sinen sin so sueptet vele waters in, so dattet te gronde sinct metten watre dattet drinct.

The nautilus, as Plinius said, is a peculiar animal that lives in the sea. At the front it has two arms, with in between – so we heard – a thin, wide membrane. If he raises his front arms high, he will sail off fast. Underneath it moves with its legs, while it steers with its tail, quite amazing. If it spots danger, it takes in huge amounts of water. And because of the water it drinks, it sinks to the bottom.

Jacob van Maerlant (Nature's digest)

Frames of reference

We come into this world as a little cloud of happiness. We are the product of highly accidental circumstances, the outcome of a sequence of events on which we did not have any influence. As our grandmother looks into the cradle she is delighted and might cry out that she looks "straight at Heaven." Maybe she hopes that things will stay like this, but that, as most of us have discovered, is not very probable. Upon delivery we are a wonderful piece of hardware and like most computers, we come with a great variety of pre-installed software. We have unwittingly been stuffed with presuppositions. At this early stage, already we have no choice other than to accept a rather arbitrary set of initial conditions and the prejudices implied by them. Later on, this collection of prejudices may turn into a handicap: inevitable but very persistent, and often very hard to overcome. For many of us overcoming them is a tough job, lifetime employment guaranteed.

There is a delightful drawing by Saul Steinberg which can serve to clarify this state of affairs. It shows the rather unique world view of somebody who was born and raised in New York – on 9th Avenue, to be precise. This New Yorker clearly has a detailed picture of his or her immediate environment, but note how the accuracy of the map rapidly decays with distance. Beyond the Hudson river we traverse New Jersey to enter a void until we hit Nebraska, pestiferous places like Las Vegas, and ultimately remote and supposedly horrid places like Siberia. The picture captures the severe

distortions of our local observer, but it also implicitly expresses a value judgment: the further away the place, the more awful it is assumed to be. Also in New York the Dutch saying holds: *"The farmer won't eat what he doesn't know."*

Staying with the Dutch for the moment: If you have been in Amsterdam, you might have seen the postcard that gives a pictorial local translation of Steinberg's drawing. No doubt there are numerous similar pictures available all over the world, which only underscores that Steinberg's theme is in some sense universal. The factual particulars of these limited views do not matter very much; the basic message concerns the relativity of such particulars. This is, by the way, also a crude pictorial version of the relativity message formulated by Albert Einstein about 70 years before Steinberg. His message was stated in a more general context, using a very different language: there is only one world unfolding in space and time, yet different observers may tell very different stories depending on their frame of reference.

In this chapter I will try to illustrate a number of interesting mechanisms that may obstruct a straightforward pursuit of the truth. The typical sequence that we encounter in these case histories is: prejudice – curiosity – discovery – opposition (through persistent prejudice or fear of the unknown) – and only sometimes acceptance. There is a price we have to pay if we keep denying new basic facts, facts that are not negotiable. In the next chapter, this cycle will be elevated to something called the "double helix of sci-

ence and technology," a rather autonomous knowledge-generating machine intimately related to the so-called scientific method. But first, let me tell you some stories.

Coding and decoding

We all have our preferred reference frame, which may or may not be optimal for interpreting a given situation. Let us stay in Amsterdam for a moment, and imagine you are a tourist doing some sightseeing. You run into various traffic signs that surprise you and you ask a local resident what the hell they are supposed to mean. She happily – and proudly – takes the time to teach the interested foreigner, and if you don't stop her it might take all afternoon.

This sign you typically encounter along the canals. It is obviously a warning to people who are swimming in the canals, to watch out for cars falling on their heads.

The second sign is typically found in the neighborhood of government buildings. It warns the employees not to leave their job early, to take their car, go home and play soccer with the kids.

The last sign speaks for itself, as it has acquired the grimmest associations worldwide early in this millennium.

Our brain is often involved in either decoding or coding activities. We all know about pictorial sequences waiting to be decrypted, very popular in intelligence tests, such as the one drawn below.

The series goes: capital M, heart, four-leaf

clover, bridge … The question is: what is the next picture in the sequence?

Now, if you have never seen this before, it may be hard to guess the next one. My experience is that even if I hand you the next entry, which is just an apple, it may still be hard to guess what is going on. It takes a little out-of-the-box thinking to break out of the misleading context that I imposed on you with my not-so-helpful descriptions. Once you see the mirror symmetry of the individual icons, the only requirement in order to crack the code is the ability to count to five!

This little puzzle illustrates the fact that problems may be hard and easy at the same time: if you don't "see it," you may find the puzzle infinitely hard and you may take forever to solve it, while if you do see it, it is little more than a triviality. It shows how relative the notions of easy and difficult are, in fact to a degree that they are not all that meaningful. Moreover, once we have seen the solution, our reference frame is changed irreversibly, which means that the sequence will stay trivial for us ever after. Decoding is basically learning, and coding is more like teaching. This is the story of science in a nutshell: *Science is decoding, technology is coding.* Science's job is to crack the cosmic code hidden in nature, and once somebody, somewhere at some instant, cracks part of that ultimate puzzle and takes the trouble to share this insight with all of us, our collective understanding has taken a step forward. Such steps are irreversible, so that the process as a whole is like the winding of a ratchet. It reminds me of the sentence Neil Armstrong solemnly spoke on July 16, 1969 while stepping out of the spacecraft to make the first human footprint on the moon (see figure on next page), *"One small step for a man, a giant leap for mankind."* [1]

That is why, although at the time of its discovery Einstein's time-dilation formula was very hard to grasp for a selected audience of the smartest people around (see figure

[1] There has been some controversy about whether Neil Armstrong actually pronounced the "a" in the first part of the sentence. I presume he did; otherwise this historical quote turns into a contradiction.

on page 21), we can nowadays teach that formula to high-school kids. Science doesn't get more complicated all the time: the core message about science is exactly the opposite, namely that it gets simpler all the time. Things become simpler because our perception is lifted to a higher and more abstract level, which is beautifully depicted in Saul Steinberg's drawing on page 22. Science amounts to simplifying the mental image we have of the world, allowing us to rationally interpret our environment and decide upon our actions.

A prosaic description of what science is, was given by a particle physicist: "We throw watches against a wall, hoping to find out how they work, and with the modern quartz watches, that is no easy task." Richard Feynman (whose phrase "the pleasure of finding things out" I have borrowed for the next section) compared practicing science with learning to play chess when you are only allowed to watch other people playing it. At some point you claim you know exactly what is going on, you understand which moves the pieces are allowed to make and what the players hope to achieve. But then somebody makes that exceptional move called *castling*, and you feel like you haven't understood a single thing. To explain what a scientist is supposed to do, I often say: imagine that you don't know anything about computers and the like, and somebody hands you the complete source code for an operating system like Vista or Mac os x and asks you to figure out what it is supposed to do and how. That is an immense challenge, virtually impossible. Anybody who has ever written an inch of computer code will know what I mean.

While science is on the decoding end of things, I mentioned before that technology is typically on the coding side. It is involved in making things, in putting things together according to certain functional rules and requirements. If you think of designing software products, then the coding aspect is very literally what technologists do. And as decoding and coding are complementary activities of the human brain, so science and technology are complementary enterprises of society. We might describe them as analytic vs. synthetic; algebraic vs. geometric; left vs. right brain hemisphere. And progress has everything to do with a successful liaison between the two.

Curiosity and the pleasure of finding things out

Fortunately, we are not at the mercy of our prejudices forever. Our most important companion in the effort to beat them is curiosity, and fortunately most of us have been generously endowed with it. We have the natural tendency to try to see beyond the tips of our noses, even to the extent that we poke our noses into other people's business. In fact, our language has an unusually large repertoire of expressions involving our noses, and that can be no accident.

However, curiosity has ambivalent connotations, one of them being that it is dangerous.

All children run into self-proclaimed authorities like parents who keep telling them in solemn voices that it is *strictly forbidden* to go through this or that door, because something horrible could happen. I remember one admonishment that said that if I were to drink coffee, I would get green feathers on my back. But we are all aware that as soon as the authority walks out, the child somehow makes its way to the forbidden door and opens it in order to find out what the hell it is that it is not supposed to know. Go ahead, child!

The attraction of forbidden knowledge is rarely fatal. Fortunately, sometimes we cannot control our curiosity. It is known that the Spanish king Philip II, one of the most orthodox and rigid Roman Catholic rulers and an

active supporter of the Spanish Inquisition in 16th-century Spain, actually kept a copy of all "forbidden" books that appeared on the *Index librorum prohibitorum* of the church in his splendid library in the Escorial palace some hundred kilometers from Madrid.[2]

The danger of knowledge is perhaps most obviously shown by the all-time practice of book burning. Books have been burned in all chapters of human history, all over the world. The picture shows "The Burning of the Books," or "St. Dominic de Guzmán and the Albigensians," a 15th-century painting by Pedro Berruguete, portraying the dispute between Saint Dominic and the Albigensians (the Cathars), who had been declared heretics by Pope Innocent III in the 13th century. The story is that the books of both the Saint and the Albigensians were thrown on a fire, Dominic's books were miraculously preserved from the flames (see the book floating high above the fire), thus proving his teachings to be true and those of the Cathars false. Today we can see the preludes to a reincarnation of such practices by agencies or regimes wanting to interfere with what we are allowed to see or read on the Internet. Censorship is a fatal surrogate for education and teaching people to make distinctions themselves.

As Francis Bacon put it: "Wonder is the seed

[2] The Index or list of forbidden books contained the names of many great writers, philosophers and scientists, including Bacon, Cervantes, Galileo, Spinoza, Marx, and de Balzac, but also Hugo, Sartre, and Gide. The list, which contained thousands of titles, expressing religious, social, or sexual morals that were incongruent with the teachings of the Roman Catholic church, was abolished in 1966 by Pope Paul VI.

of knowledge." How then is it possible to have expressions like "Curiosity killed the cat"? Why is it that curiosity has so many negative connotations in daily life, while it is clearly such a crucial driving force behind progress, scientific progress in particular? Most parents are familiar with the inquisitive child who pushes the intellectual quest till the bitter end: "But why is that so?" The parent, slowly losing ground to their beloved's persistent questioning which peels off layer after layer of their limited knowledge till their ignorance is exposed, throws both hands in the air in despair and screams: "It is that way because that's the way it is, okay!" By this time the parent has been completely defeated, and the brat may triumphantly start reciting the nursery rhyme: "Why? That's why! 'That's why' is no reason!"

We are dealing here with what in the fancy jargon of science policymakers is nowadays referred to as *curiosity-driven* research. This process which starts from wonder raised by simple observation, from ordinary questions about simple facts, has ultimately led to the most astonishing insights into the workings of our world.

When my eldest daughter was about six, she asked me what a triangle was. I explained it to her and she found it quite obvious. Next she asked about a quadrangle … Oh yes, of course! And could I also inform her about a *two-angle* and a *one-angle*?[3] The *zero-angle* was again obvious for her: that could only be a circle. Actually, this was quite a heroic conversation on her side: a mighty combination of observation and imagination, combined with a strong urge to get to the bottom of things. Another scientific debate with one of my children revolved about the question of what constituted a "table of two," in particular the question whether the sequence of numbers 1, 3, 5, 7, … was such a table or not. In fact they went one step further by asking why the sequence a, c, e, g, … was not such a table. After all, there too you had to move forward two positions at a time, or alternatively, you had to skip one entry each time. What would you have answered? What answer could you have come up with if you would by accident have forgotten about the key difference between multiplication and addition?

Painful though it is to admit, I think the problems start right there, if we as parents fail to be alert. When our children force our hand with some crystal-clear question to which we do not know the answer, we tend to fob them off – instead of rewarding them with plenty

of attention. After all, that is exactly what we do when we go and listen to their musical performances, or when we stand around a sports field on a rainy Sunday afternoon, screaming our lungs out to encourage our loved ones who are running after a ball.

It is so easy as an educator/parent to let the analytical gifts and the talent for abstraction slip through our fingers. In that respect we get what we deserve by the time

[3] We came to some conclusions on the subject: A two-angle automatically turned into a four- or three- angle in the plane. However, on a sphere it was possible to draw an equilateral two- angle. A one- angle in the plane would be an infinite wedge.

Norman
Rockwell

KNOWLEDGE IS POWER
Post Cover • October 27, 1917
69

critical investigators and analysts? Einstein once remarked that it is a miracle that curiosity survives formal education Maybe if we compare the schools depicted by Jan Steen (page 24) and Norman Rockwell (this page) we should prefer the former after all! Steinberg showed us the way implicitly in the pictures I showed at the beginning of this chapter: even a New Yorker needs to travel to extend his vision and let it become more realistic. Facts are resilient, and new facts are the worst in this respect; they may haunt us and ultimately force us to radically revise our most basic beliefs.

According to reliable sources …

One of the unexpected threats to learning the truth, or even of just fact finding, is the omnipresence of authority. Experts are always ready to provide us with an overdose of unsolicited (and in hindsight often useless) advice. You know how it feels when people are already throwing answers your way before you are even allowed to formulate a question. Answers to questions you never asked are a pest. But if you overcome authority and curiosity does drive you into new territories,

these youngsters knock at the gates of our Universities, which traditionally take pride in the fact that they allow access only to the very best and brightest, in particular in the natural sciences. It appears that we teach our children how to consume rather than how to think and create. Why do we teach them facts instead of teaching them how to observe, to wonder and to be creative? Why do we turn them into believers rather than

then you are faced with new questions: "How do we know what to believe?," and: "How reliable are our observations?"

As a young child, like most children in the Netherlands I believed in Saint Nicolas, a very old, good-natured, silver-bearded fig-

⁴ There is an evolutionary tree of Santa Claus myths tracing them back to the 4th-century benevolent bishop Saint Nicholas of Smyrna (now situated in Turkey). The feast of Saint Nicolas I describe is still celebrated on the 5ᵗʰ of December in the low countries in the North-West of Europe, and it remains pretty close to the original celebrations. Saint Nicolas is quite different from the jolly good old fellow in the sleigh with the raindeer associated with Christmas whom we find in the Anglo-Saxon version, developed in the 19th century.

ure who would bring me loads of presents.⁴ I was a strong believer, as I believed what my parents told me, and after all I had seen the bishop in person arriving on an old steamboat in the harbor of Amsterdam and cruising through the shopping mall. In the night he supposedly rode over the roofs of the houses on a white horse, while his servant Black Peter would throw presents through the chimney. Somehow these ended up in our shoes, which had been posted in front of the fireplace or stove. In our shoes we put a carrot or some straw for the horse, and a drawing for the Saint to make sure that he would be pleased with us. There was no reason whatsoever to doubt his existence or the stories about him: we had seen him riding his horse, and the presents that showed up in our shoes in the morning were real all right.

The origins of this legend are quite amusing, as it is a peculiar mixture of hedonistic and Christian customs that evolved into a rather bizarre educational tool over time. Saint Nicolas used to carry a heavy bible-like book in which he kept a detailed record of all the good and bad things that every child had been doing over the year. Indeed, that was the first "Big-Brother-is-watching-you" type of experience I had. If you had been a nice kid all year, you would get sweets, but if you had been naughty, Black Peter would put you in a bag and the bishop would take you away to his palace in Spain. Presumably he had much help from his servants, who were continuously spying on our behavior, pretty much like the Russian KGB in its heydays. Feeble educators would often abuse this story to threaten kids with "the bag" in case they wouldn't obey. Painful and remarkable is also the racist undertone, as the Saint has these black helpers who traditionally are portrayed as not being very much on top of things. Even though attempts have been made to introduce green, red and purple Peters, the prejudices have never been officially corrected, because they form part of a cultural heritage. I'm afraid that in many schools in the low countries children may still sing songs with texts like: "In spite of being black, his intentions are good." Another somewhat shady association, which already caught the attention of quite a few psycho-analysts decades ago, was caused by the

common pictures in children's books of the old prelate with a small child on his lap, that begged for a Freudian interpretation.

As a kid I took part in these profitable celebrations as a true believer, until I finally began to wonder how all this could be possible and my devotion started to show some cracks. There were certain observations that didn't help: I once saw two Saint Nicolases at the same time in the same street, at another instant I was quite sure to recognize the voice of my uncle, and one day I found the closet of my parents' bedroom stuffed with presents. But beliefs are persistent, and I was all too eager to accept any explanation that my parents were equally eager to offer so that I didn't have to give up my cherished belief.

The new rank of *Assistant Saint Nicolas* was introduced, as one bishop could obviously not carry all the presents for all the children, and that was of course also the reason why he stored them in certain places like parents' bedroom closets.

But at a certain point curiosity prevailed, and I had to come to terms with the plain fact that the whole thing was a fraud – I think I saw him taking off his beard! What? Had my parents been deceiving me for years? Yes indeed! After admitting this, my parents immediately invited me to join the crowd of the initiated and lie to my younger brothers about the whole thing. If you can't beat them join them! The idea of treason did not occur to me, because in accordance with my new status I was advised to never take somebody's belief away. That was considered to be something like stealing. There was another price to pay for not believing: I was no longer eligible for putting my shoe in front of the fireplace. No belief, no presents.

The whole Saint Nicolas experience had great educational value: it taught me to not always believe right away what I saw or what I was told to be true. After all, how could Black

Peter ever climb through the chimney and deliver the presents neatly in my shoe? More interestingly, I had to come to terms with the painful discovery that the greatest authorities and most reliable allies in the world, namely my own parents, did not always tell the truth! Looking back, it showed me how hard it is to resist the lure of becoming part of the "establishment" and adopting its deplorable habit of cheating. These lessons proved valuable in my later life as a scientist, where being critical in accepting evidence was crucial. I tell you this little story because it has all the basic ingredients of the value of curiosity and shows how the hardship of demystification is unavoidable. Nobody is beyond reasonable doubt … not even Saint Nicolas!

Crossing borders

We like to travel to learn about other parts of the world and to broaden our view. As a record of our travels we write letters or emails, and put together blogs and photo albums. The medieval poet Jacob van Maerlant, who lived in Flanders in the 13th century, wrote a remarkable pedagogical encyclopedia of nature called *Der naturen bloeme* (Nature's digest). He modeled it after *De natura rerum* (On the nature of things) by a learned Dominican priest named Thomas de Cantimpré (1201-1277), who in turn took his inspiration from the work with the same title by the Roman poet and natural philosopher Titus Lucretius Carus, who lived in the first century BC. Van Maerlant's work counted no less than thirteen volumes, about plants, animals – including a magnificent collection of sea monsters, one of which is shown at the beginning of this chapter – and the human species. It was a very large knowledge base, certainly for that time, and an especially striking fact was that it was not written in Latin but in the ordinary Flemish language. Van Maerlant was a master in science popularization and outreach before such a thing was even invented. In hindsight we can see that this huge collection of data actually contains the strangest entries, topped with an icing of suspense that sparked the imagination. It may have caused quite some sleepless nights in medieval Flanders. The balance between fact and fiction frequently tipped over to the fiction side. Take for example Van Maerlant's account of how people supposedly behaved on the other side of the globe (see picture on page 28):

Other people are there, who live from the fragrance of apples, without need for other food. If they have to travel large distances, they keep an apple in front of them, because they are bound to die if a bad odor would reach them.

You see, vegetarians and even vegans still have a long road ahead of them! And what about other parts of the world, where the hair of women turned black instantly after giving birth to their first child? Or the people where the women only gave birth to quintuplets (see picture on the previous page)! These are no doubt awkward circumstances, but still highly preferable over the remarkable people who had the habit to beat their old folk to death once they were "worn out," and after that would sit down together and eat them. Van Maerlant observes soberly: "And this they consider a benefaction, it neither raises guilt nor is called evil." We might frown or smile about such ignorance, but it reflects the simple fact that in the early Middle Ages planet Earth was still, to a large extent, *terra incognita*. When you think about it, these can be seen as an early version of Steinberg's picture, and I doubt whether the present switch to a global tool like Google Earth would make an essential difference.

We are clearly dealing with a very exciting but equally unreliable account by a learned man. In spite of his undoubtedly laudable aspirations to teach the people true facts of nature, he only succeeded to a limited extent, but Van Maerlant certainly raised a lot of curiosity with his disclosures. The situation nowadays is more similar than you might be inclined to think. Now too it is often a small step from hearsay to "from a reliable

source we learned …" We had better keep asking ourselves how reliable our sources really are, although facts allow for multiple interpretations as illustrated by the painting called *The Gardener* by the Italian painter Arcimbolo (1527-1593).

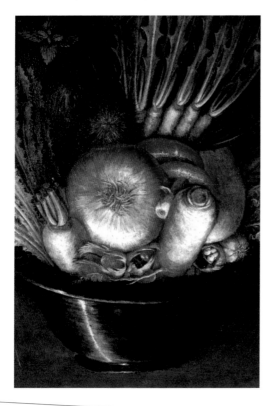

I must admit that I myself do not do much better than Van Maerlant. I like to take pictures on my travels, and have the habit of spending time behind my laptop – even before I get home – to create a representative album of my adventures. Generally I do such a thorough job of cutting, pasting and improving that the thrilling end product is barely representative of what really happened on my trip. The term "retouching" captures very well this process of rewriting and brushing up my personal history. Maybe you recognize what I am talking about: photographs in which you look clumsy are eliminated; other images are cut in half because this "improves the composition." Ugly views from hotel rooms? Never heard of. In the long term, more radical surgery is applied to the albums. Certain pictures are removed when they no longer fit the occasion: the ones in which you are enthusiastically smoking pot, unwanted ex-partners are visible, and so on. What we end up with is a plain counterfeit of our own past, presumably because of our strong need for perfect memories.

This form of permanently face-lifting our own past turns life into a kind of never-ending job interview. The Croatian author Dubrovka Ugresic has described this as a process in which each of us is busy erecting monuments for ourselves. And like other monuments – generals on horses, liberators of the people, tombs, mausoleums – our innocent ego-monuments look an awful lot like each other. Photo albums share the kind of universality and relativity of the Steinberg pictures mentioned earlier. And with sophisticated tools like *Photoshop*, the "eternalization" of our splendid youths will keep on converging towards a single most beautiful representative: "Everybody happy? Yeah!"

In the Internet age, this story continues with personal home pages and pages on business and social network sites like *LinkedIn* and *Facebook*, where you can post a jazzed-up profile matching your desires rather than the bare facts. In fact, the profile may not even reflect your own desires, but rather the expectations of your social surroundings.

The price we pay for this conformism is a dramatic collapse of diversity. It resembles the public obsession with particular brands of clothing, which similarly is an assault on the splendid diversity of personal taste. In these instances majority rule and common expectations replace individual judgment, and "knowing what to wear" takes precedence over authenticity. All these tendencies are desperate attempts to bootstrap oneself up into the dream world of the ideal consumer.

You might find the tone of my discourse somewhat lighthearted, while in fact we are dealing with rather serious matters. Obviously, I am aware that there are far more dramatic examples of misrepresenting the facts. The construction of artificial truth by systematically cutting and pasting pieces of objective evidence is nothing new. Political systems have exploited this trick of selecting and reassembling the facts for ages, to justify ludicrous actions, such as burning books, locking up innocent people, and fighting what later turned out to be avoidable wars. We all know how world powers tend to rewrite history until the historical necessity of their interventions and dominance becomes self-evident. Nowadays this is achieved through e.g. grossly misleading press releases, severely restricting Internet access, house arrests, and extensive spying systems. "History is written by the victors," is an old saying. Even so, often there are many versions of history accounting for the same sequence of events.[5]

How can we avoid this gross deformation of the truth caused by selecting only those parts of the truth which suit our needs? In absolute terms this will be hard, because a finite number of observations always allows for a multitude of "truths" fitting the data. If, however, we are not after the absolute truth, just a *better* one, then the method to systematically improve our knowledge is surprisingly simple and has been well-established in the scientific world for ages. To significantly raise the truth value of observations, one should combine data from several independent observers. Add the accounts of other travelers who visited the same place. Let us compare the notes of the

policemen to the account of the students who were the instigators of the upheaval that got out of hand, though they were merely striving for a better world. And don't forget to add the observations of some outsiders who happened to pass by when the riots took place. In general, the idea is to guarantee that there is no monopoly on providing evidence and information – rather the opposite. If the facts from different sources appear to be inconsistent, and we trust the sources, then we are compelled to enlarge the framework in which we interpret and discuss these facts. It means we have to transcend our old points of view, because true facts are never inconsistent. This transcending plays a key role with respect to the notion of turning points in science, as discussed in this book. Good science is truly democratic in that it allows many voices; it also allows grave mistakes, because those usually accelerate progress.

The persistence of prejudice

Only I must observe that the common people conceive those quantities under

[5] For example, in May 2009 president Medvedev of Russia announced that he had appointed a committee to start rewriting Russian history, aiming to restore the heroic roles of the totalitarian leaders (like Stalin) of the communist era.

no other notions but from the relation they bear to sensible objects. And thence arise certain prejudices, for the removing of which it will be convenient to distinguish them into absolute and relative, true and apparent, mathematical and common.

Isaac Newton

Science has nothing to do with [religion] except in so far as the habit of scientific research makes a man cautious in admitting evidence.

Charles Darwin

I have argued that obtaining reliable information requires independent observations and the freedom to report about them. This implies that the independence of agents who deal with information, be they politicians, judges, journalists or scientists, should be guaranteed in the constitution. In Western society we are careful to separate church and state, and the legislative and executive branches. Yet in the world of media and science we appear to be much more lenient with respect to the direct influence of market forces, which are known to not always respect the full truth when profit comes into play. Independent research means that integrity should be rigorously protected in all realms of a human research effort.

Science requires democracy and transparency in its procedures, but, surprisingly, is rather undemocratic in its findings. Scientific truth is not valued by majority rule; it bears no relation to what we *want*. In science a single man can be right while everybody else is wrong. Testing a scientific truth has to do with carefully designed experiments, not with a popularity contest. In that sense the disparity of success in the popular media and politics and success in science could not be larger.

The reliability of observations is indispensable for science, and the simplest way to ensure it is to demand that measurements should be *reproducible*, in other places and at other times. Now, amusingly, there used to be a curious journal with the title *The journal of irreproducible results*, which one might easily imagine to be the fattest of all scientific journals. Even so, it would still be surpassed by the Internet, which undoubtedly hosts the largest collection of irreproducible facts. The journal deserves credit, however, for the fact that its name bears witness to a clear measure of integrity. It points to the problem of obtaining unequivocal evidence even in the empirical sciences. Theoretical scientists with some irony often say things like: "Don't believe experiment until it has been verified by theory." The complicated relationship between observations, imagination and interpretation remains a delicate issue even in good science. Scientists are just like people, you know, and like most people they have their very personal convictions, their heroes and their hunger for fame. The latter is not hard to understand if you know how much at present their funding depends on it. This state of affairs may jeopardize scientific integrity.

Even if they do not lust for power or fame, scientists can have a hard time dealing with the hard facts. They have their prejudices, like all of us, but the factual outcomes of their research don't seem to care very much about their preferences and expectations. Real facts are non-negotiable. It is illuminating to briefly contemplate how even the greatest minds had a hard time accepting the conse-

quences of their own brightest discoveries. The reason is that these discoveries tore down some of the most cherished pillars of established wisdom – or rather, lore.

There is a well-known anecdote told by the famous physicist Victor Weisskopf, about the lectures on relativity and astrophysics which he attended as a student in Göttingen before the Second World War. The lectures were given by Monseigneur Lemaitre, a Belgian priest. This researcher made absolutely brilliant fundamental contributions to Einstein's General Theory of Relativity, notably concerning the mathematics of the celebrated Big Bang model – which describes the evolution of our universe in great detail. In those days, Lemaitre was particularly interested in determining the age of the earth, using a dating method based on the abundance of long-living radioactive elements in the crust of the planet. Nowadays this radioactive dating method is the bread and butter of many fields: it is used on a large scale to determine the age of sediments, of the remains of our predecessors, and of paintings that are attributed to old masters for dubious reasons. Weisskopf recalls the following conversation

36

after a lecture in which Lemaitre had calculated that the earth ought to be about four and a half billion (4.5 x 10⁹) years old:[6]

When we were sitting with him after his talk, someone asked him whether he believed in the Bible. He said, "Yes, every word is true." But, we continued, how could he tell us the earth is 4.5 billion years old, if the Bible says it is about 5,800 years old? He said, I suppose tongue-in-cheek, "That is no contradiction." "How come?" we nearly shouted. He explained that God had created the earth 5,800 years ago with all the radioactive substances, the fossils, and other indications of an older age. He did this to tempt humankind and to test its belief in the Bible. Then we asked, "Why are you so interested in finding out the age of the earth if it is not the actual age?" And he answered, "Just to convince myself that God did not make a single mistake."

[6] Over time, the age of the earth has led to numerous controversies. Creationist estimates based on the Bible yield ages of less than five thousand years. This is by the way more than the Mayan estimate. In fact Lemaitre's point of view was somewhat anachronistic, if you think of the fact that Philip Henry Gosse in his *Omphalos, an attempt to untie the geological knot* of 1857 already pointed out that taking the biblical creation story seriously implied the negation of hard geological facts. Darwin estimated the age to be 300 million years, based on geological evidence, and ran into fierce opposition from the great physicist Lord Kelvin, who until his death maintained that its age was no more than 30 million years. His calculation, we now know, left out part of the relevant physics. The modern determination, obtained by applying radioactive dating methods to the oldest rocks and to a large body of meteorites, yields an age of 4.55 billion years (with an uncertainty of about one percent), so the present situation basically confirms Lemaitre's result. Incidentally, this is also considered to be the age of the solar system as a whole.

Knowing the facts is one thing, but accepting them is clearly something else. This reminds me of what I told you about my experiences with Saint Nicolas, and it is the kind of story that keeps repeating itself in the history of science. The great physicist Hendrik Antoon Lorentz took active part in the creation of the Special Theory of Relativity, but meanwhile he could never really accept Einstein's fundamental postulates that abolished the *ether*.

This ether was believed to be an all-pervading substance, needed for the propagation of electromagnetic waves such as light and radio waves. Einstein showed that it was not necessary and that those waves could propagate through empty space. But Einstein himself has been known to resist evidence, too. The most far-reaching consequence of his General Theory of Relativity was that it implied a dynamical universe – a universe that is not eternal and static, but one that evolves in time according to Einstein's celebrated equations. Surprisingly, Einstein was strongly opposed to this idea, at least in the beginning. Later, after Edwin Hubble discovered that our universe is indeed expanding in 1928, Einstein called his earlier resistance against the idea "the greatest blunder of my life."

Another story about Einstein concerns quantum theory. The scientific giant who conceived relativity also stood at the cradle of quantum theory; strangely, he received the Nobel Prize for his contributions to the latter – the photoelectric effect, to be precise – and not for his magnificent achievement of relativity. Such is the relativity of Nobel prizes. In spite of the tremendous successes of quantum theory, Einstein refused to accept the basic axioms of the theory. The radical change to a fundamental uncertainty and a probabilistic interpretation of physical reality, which was enforced by quantum theory, was insurmountable for him, and brought him to his famous exclamation that it couldn't be true, because "God doesn't play dice."

That brings us to Erwin Schrödinger, who was the first to write down the fundamental equation that decoded the quantum nature of reality. He detached himself to a certain extent from his brainchild once he realized how radical the conceptual revolution was that it had started.

A final example concerns the famous British physicist Paul Dirac, who reconciled quantum theory and special relativity in the beautiful equation named after him. Apparently he was not amused when it became clear that that equation implied the existence of *anti-matter*; the unheard-of outcome that for each particle type in nature there exists a counterpart with the same mass but with exactly opposite other properties such as charge. Of course Dirac was quite happy after all when antimatter showed up in experiments a few years later.

I tell these stories to show how hard it may be to accept the consequences even of your own discoveries. This happens especially if the outcomes go strongly against the pre-conceptions or expectations you cherish. It can be motivated by fear to unleash too big a scientific revolution. Such an overdose of impact has knocked even giants from their feet, while in hindsight these profoundly new perspectives were exactly what made their contributions so outstanding in the evolution of science.

Unveiling Mother Earth

The Greeks already figured out that the earth was a sphere. They assumed it was at rest in the center of the cosmos, and Eratosthenes even determined the circum-ference of the earth to be 252,000 stades, which amounts to about 45,000 kilometers, about 15 percent off from the correct value. In the following centuries the Romans main-tained the Greek view. Astonishingly, this view got lost in the early Middle Ages, when apparently a widely held view was that the earth was flat. The big question was wheth-er that flat world was finite or infinite. Was there an edge, where you could fall off the earth into an abyss of the unknown, which according to the experts looked pretty much like hell? Of course this fate could only hit you if you would be so imprudent to go far away from home and search for that edge. The second possibility was an infinite plane. From a strictly scientific point of view, the flat-earth hypothesis is quite respectable as a hypothesis, because it is falsifiable – it can be proven wrong by careful observa-tion. Indeed there were known facts that clashed with a flat earth. It was known that ships that sailed far away disappeared un-der the horizon; that one could look further

if one stood on a highly elevated point; and that during a moon eclipse one could observe the disc-shaped shadow of the earth. Such facts could easily be confirmed by any individual wishing to do so, and these facts could therefore not be swept aside easily. They became a decisive factor in the debate, and the flat-earth point of view eroded away over time. This episode illustrates another rule cherished in good science, namely that a wrong theory is far more valuable than a vague theory. As Richard Feynman once put it: "We are trying to prove ourselves wrong as quickly as possible, because only that way can we find progress."

Theories that cannot be proven wrong may make an interesting story, but scientifically they are a pain. Lacking a systematic elimination of failing alternatives we end up with a *status quo* of opinions, which frustrates progress and frequently deteriorates into a futile fight among factions sticking dogmatically to their own views. In extreme cases, students may even be told to not read any papers written by the opponents, which goes very much against the spirit of scientific research.

We must be grateful to courageous explorers like Christopher Columbus, who were the first to put the spherical-earth hypothesis to a real test when they kept sailing West, hoping to find an alternative route to the far East. We all know that they never made it there, but discovered America instead – a splendid example of *serendipity*, making an accidental discovery that has nothing to do with your original research goals.[7] As a matter of

---

[7] The comparison between scientific discovery and the expeditions of explorers can be extended. Victor Weisskopf compared today's engineers with the shipbuilders, without whom the enterprise could never have started. Experimental physicists are like the explorers making the

trip: they were the first ones to see the new land and leave their footprints there. The theorists he compared with the people who stayed behind in Lisbon and predicted that Columbus would end up in the East Indies. Sometimes a continent lies between theory and experiment…

fact the great explorers were courageous indeed, because even on a spherical earth you could drop into a hole or slip off the side. And if you did not slip off, you might have to deal with the species of the Antipodes, living on the southern hemisphere (as their name suggests). The fact that the antipodes didn't fall off the earth implied that they had to be quite different from normal people, certainly inferior and probably very nasty, as depicted for example in the *Margarita Philosophica* from 1517 (see picture on the right).

It must not have been easy to travel around the world with in the back of your head the suggestive geography of the world and its many heavens of Dante's *Divina Commedia*. One could easily end up in Hell, which was located somewhere in the deep south, as depicted much later by Michelangelo Caetani in *La Materia della Divina Comedia di Dante Alighieri* (see picture on the following page).

The recognition of living on a curved surface was revolutionary. It took not so much "outside the box" thinking as "outside the plane" thinking. It is exactly as Einstein once remarked: "The significant problems we face today cannot be solved at the same level of thinking we were at when we created them." Living on earth we are confined to the two-dimensional surface of a sphere. Such a surface has two very special properties: it is without boundary (in the sense that it has no edges) and yet it has a finite surface area. Without boundary but finite may sound contradictory, but this insight into our situation leads to important social and cultural consequences. A finite area can be fully explored; apparently we were living in a knowable

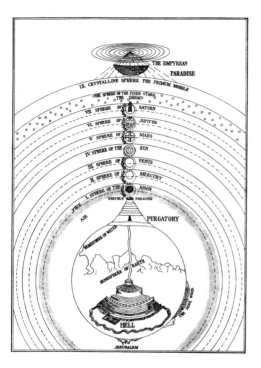

is the only inhabitable planet we know, the bumper sticker slogan "Love it or Leave it!" does not apply – at least for the moment. The French Jesuit and paleontologist Teilhard de Chardin (1881-1955) appreciated these features in the 1940s. In *La phénomène humaine* he argued that a social and political integration on a worldwide scale was inescapable, proving himself a true globalist.

This feature of the earth having no boundary shows how a deep mathematical theorem from geometry can have profound consequences. The theorem amounts to the simple statement that a boundary itself has no boundary. The earth surface is the boundary of a solid sphere, and therefore that surface itself has no boundary. As a consequence of our modern spatial awareness, we are accustomed to thinking of the earth's surface as the surface of a ball, which we envisage embedded in the ordinary three-dimensional Euclidean space. In this system a point is labeled by three numbers: length, width and height. This embedding picture is not necessary, however.

The surface itself is two-dimensional, and to describe it, it is sufficient to draw two-

world. In our era we have been confronted once more with an important consequence of this fact: we have come to realize that the earth's resources are finite, and this implies that there is a limit to economic growth. Mankind as a whole should work towards a sustainable world, based on cyclical principles rather than unlimited expansion. As this dimensional maps which can be collected in an atlas. This collection of flat pictures allows us to reconstruct all the properties of the surface, including its topology[8] and curvature. To learn about the topology and the curvature properties we do not have to leave the surface at all: it suffices to do strictly two-dimensional experiments. Suppose a certain Ms. Ant is standing on the surface of a basketball and starts walking "straight ahead" in some direction along the surface. If she keeps going, she will return to her point of departure. Without ever leaving the surface she will have proven that she is not living on a flat infinite surface. On the same basketball, Ms. Ant could also draw a large triangle, made up of straight lines on the surface. If she would measure the three angles at the corners, upon adding them she would find – time and again – that the sum would exceed 180 degrees. Quite a shock for Ms.

Ant, who took two years of geometry in an ordinary high school! Again the lesson is: when you start measuring, you don't always find what you expect.

I will conclude this section by mentioning that the discovery of the geometric structure of the earth was the very first time that mankind had to exchange the notion of a flat space for a curved one. But not the last one. Further on, I will discuss in more detail how with the advent of Einstein's General Theory of Relativity we had to add one more dimen-

[8] The topology of a closed two-dimensional surface is determined by the number of holes in it. The sphere has zero holes and is therefore topologically equivalent to a table. A donut has one hole and is for example topologically equivalent to a coffee mug with a handle.

sion for the universe as a whole. Instead of speaking about a circle as a closed curve in the plane, or a spherical surface as a two-dimensional curved space, scientists had to start thinking about a three-dimensional spherical "surface" as a hyper-surface forming the boundary of a four-dimensional ball; a three-dimensional space that itself has no boundaries and a finite volume. As was to be expected, the suggestion that our world is part of a curved spacetime was met with resistance and disbelief, all the more so be-

cause the geometry of a three-dimensional spherical surface goes beyond direct human experience and imagination. We can only understand it by employing formal analogies or by employing a certain non-Euclidean geometry that was developed by Riemann in the nineteenth century. Apparently we need all our imagination and the creation of new languages to accurately describe the way things really are.

Thunder, lightning and black holes

New scientific insights very often generate firm resistance. This is partly due to the healthy criticism of skeptical individuals, but another part is due to fear, fear of losing ground. The plain fact that there exists a convincing body of observational evidence doesn't make much of a difference to many people. Imagination and expectations overrule observations, especially if these lead to an inconvenient truth. And for many people a myth is just more attractive, if only because it offers more options for interpretation, which makes it more flexible and thus useful. Yet this preference for myths in areas like health or economics has often led to painful delu-

sions. And people who exploit the innocence and good intentions of the public may in the end be led to not-so-innocent fraud.[9]

Consider the rather dangerous experiments Benjamin Franklin together with his son William conducted on the banks of the Schuylkill River near Philadelphia in 1752. He went out during a heavy thunderstorm to fly a kite, hoping that a lightning bolt would strike the poor kite. This would allow him to prove the conjecture that lightning was just an electrical discharge from a cloud to the surface of the earth.[10] If this was true, the wire of the kite would guide a tremendous current towards the earth, and that would not be hard to measure by some device. I suspect that the picture is not accurate, because using your hand to measure a current is clearly not the way to do it. In fact there

is a story about a Russian contemporary of Franklin, a certain professor Richman, who conducted a similar experiment near St Petersburg. Out of curiosity he bent over too close to the wire when it got hit by the lightning, so that a fireball leapt over to his head and killed him. Which illustrates that curiosity is not *always* a blessing. Sure enough, Franklin did succeed and happily announced his findings.

In the old days, natural phenomena like lightning, floods, forest fires, and earthquakes were believed to be the result of supernatural interventions. They were considered to be signs of some divine dismay that had to be interpreted. Indeed most religions had

[9] See for example *The economics of innocent fraud* by John Kenneth Galbraith, 2006.

[10] Strictly speaking, it sufficed to show clouds would be charged with respect to the earth, so that a current would already flow by having the kite in the cloud. Storm clouds are negatively charged at the bottom and positively charged on top, explaining the picture of lightning on the right.

FIG. 160. —FRANKLIN'S EXPERIMENT.

somebody had sinned, and somebody had to be blamed. The priests, as the main interpreters of the bad news, gained more control over their communities that way.

Soon after his experiments Franklin turned his discovery into a very simple but immensely useful invention: the lightning rod. One puts a copper rod on top of a roof, with a conductive wire leading to the ground, and is forever freed from the divine fury. It worked better than buying indulgences or saying prayers. This brilliant device found its way to the public – except of course to those who would optimally profit from it, the church fathers. In his two-volume *A History of the Warfare of Science with Theology in Christendom*, Andrew Dickson White gives an account of how the Roman Catholic church finally became convinced after a disastrous accident took place in Brescia. The Venetian republic had stored some two hundred thousand pounds of powder in the vaults of the San Nazaro church. In 1769, seventeen years after Franklin's discovery, a lightning bolt hit the church and the powder exploded. One sixth of the city was destroyed and more than three thousand people died. It took until

built rather extensive theological doctrines around these natural catastrophes, which often involved some Supreme Being getting upset or angry about some human action down below and you had better not interfere with it. Needless to say, this supreme discontent bred fear among the people. Apparently

1777 before a *pagan rod* was finally placed on the cathedral of Sienna. Then the debate seemed settled, at least in Italy.

You might object that this story is a bit stale, as it took place so long ago. I would counter that it is exactly the distance in time which makes the absurdity of the situation clearly visible and understandable. A striking aside is that even nowadays these things still happen, as the incident reported in 2003 in *The Findlay Courier* shows:

> *A guest evangelist at First Baptist Church in Forest, Ohio was preaching about penance and asking for a sign on Tuesday night, when the church's steeple was hit by lightning, setting the church on fire. "It was awesome, just awesome," said church member Ronnie Cheney. "He was asking for a sign and he got one." About 7:45 p.m., lightning hit the First Baptist Church's steeple, went through the electrical wiring and blew out the church's sound system. Cheney said the lightning traveled through the microphone and enveloped the preacher, but he was not injured. Afterward, services resumed for about* 20 *minutes, but then the congregation realized that the church was on fire and the building was evacuated. There were no injuries. Damage to the church was estimated at about $ 20,000.*

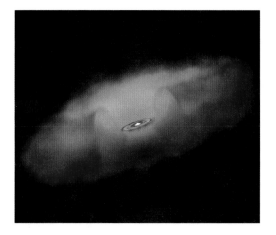

Hilarious as this may sound to us, we also know that in many places in the world, natural disasters leading to far more serious situations are still interpreted as punishments from a Supreme Being. I myself remember how when the dykes broke and a big flood came over the South-West of the Netherlands in February 1953, causing almost two thousand people to drown, many concluded that our God-fearing nation had been punished.

Nowadays we have a contemporary variation on the "pagan rod." In fact it is almost an inversion, because this time science was ac-cused of causing disasters instead of help-ing to eliminate them. I am referring to the conceived danger of turning on the largest particle accelerator ever built, the Large Hadron Collider (LHC) at the European re-search center CERN in Geneva. The discus-sions could have resulted in a great delay, and not because of technical problems such as leaking vacuum chambers, overheating power supplies, quenching superconduct-ing magnets, shortcuts in detector circuitry, or failing data-collecting software.[11] No, the objection, threatening to lead to years of delay, came from a rather unexpected source. It exploited not so much the persis-tence of prejudice as the layman's fear of the unknown. In March 2008 Walter Wagner, a retired radiation safety expert from Hawaii, and Luis Sancho, a Spanish science writer, filed a lawsuit in Honolulu contending that the collisions between particles in the LHC would cause the release of such enormous

[11] There has been such a delay starting October 2008 be-cause of substantial damage to the ring, caused by a tiny but quite fatal soldering defect in one of 10,000 connec-tions.

energies in such small volumes that microscopic *black holes* could be created. These could end up swallowing the entire earth. A picture of a black hole is per definition not of the black hole itself but of the things that may happen in its surroundings, like the formation of an accretion disk (page 47). Also the production of dangerous particle species known as *strangelets* – a sort of contagious dead matter – or so-called *magnetic monopoles*, which could catalyze the total destruction of ordinary matter, would pose a threat to global safety.

The two men demanded an instant stop of the construction work on the collider, until it was proven to be safe. Here was an eight billion dollar machine on the verge of completion, and some radiation safety expert from Hawaii was trying to block the whole thing! If they were right the machine and its gigantic detectors would be the first things to disappear. In the figure on page 46 you see the the ATLAS detector in an early stage of construction. The scientists who were considered to be the real experts basically all agreed that such fears were pure fantasies, and that this lawsuit was nonsensical.

The arguments were based on faulty estimates using highly speculative theoretical ideas. But the problem with highly speculative theories is precisely that although you may not be able to prove that *something* will happen, you can't prove that this *something* will definitely *not* happen either. If moreover this *something* is a total doom scenario like the one proposed, in which the whole Earth would be gobbled up by a black hole created by some super-smart nerd collective called "particle physicists" one hundred meters under the earth's surface near the nice city of Geneva on the French-Swiss border, people are quite likely to take it seriously. Nothing is as easy as creating fear with fiction, and fear is often contagious. It spreads like a virus, and once that has happened, it takes far more than reasonable arguments to remove it from the heads of those who have been infected by it. Even when experts get together and give the verdict *Safe, don't worry, be happy*, there is still the possibility – which is hard to exclude logically – that the experts are all part of some hidden conspiracy. A nice follow-up of Dan Brown's *Angels and Demons*, demonstrating how reality can be taken over by imagination.

The end of the LHC story gave some hope to the scientific community, but that will not keep others from attacking it. A group of independent experts, not involved in the construction or any of the experiments, was invited to prepare a safety assessment of the LHC with respect to the danger of producing certain high-energy states of matter.

The resulting report was firm in its conclusion: if any of the proclaimed dangers would imply a realistic threat we would have known about it for a long time, because many tremendous ex- and implosions would have been observed in nearby parts of the universe, due to the impact of extremely-high-energy cosmic rays and particles, exceeding many times the energy of the particles in the LHC. Unusually strong committee language saved the day: "The Universe as a whole conducts more than ten million million LHC-like experiments per second. The possibility of any dangerous consequences contradicts what astronomers see – stars and galaxies still exist." Thanks to this information, the US court dealing with the case could take action and dismiss the case, stating that it was "overly speculative and not credible." And so this "Doomsday suit," intended to halt the world's largest particle collider, was thrown out of court.

From evolution to HIV

Another debate that almost acquired a scientific aura took place in the Western world around the turn of the current century. A religiously-motivated issue attracted considerable attention, because some important religious and political leaders took firm stands in it. The debate revolved around conceptual issues connected with the theory of evolution. The "intelligent design" movement took the floor with the claim that there are clear examples of "irreducible complexity" in nature. Their claim was that the central paradigm of modern biology, namely that all organisms have evolved from the most primitive life forms according to the principles of Darwin's evolution theory, failed to explain some examples. The existence of these examples was said to render the theory of evolution untenable.

This attack did not take place according to common scientific practice, which requires

that one submits the results of carefully conducted research to reliable peer-reviewed journals. The more important the issue, the more prestigious and critical the selected journal should be. However, the intelligent design movement's attack strategy included a familiar motivation for deviating from standard scientific practice. It boils down to the remark that the scientific establishment would never allow the publication of a paper that overthrows a central paradigm. No wonder: nobody is going to provide support to the bombing of his or her own house! Many scientists agree that this strategy, echoing the rhetoric of pseudoscience, is not very convincing. The system for selecting papers is far from perfect, but it has at least been fair enough to allow many revolutionary breakthroughs, such as the discovery of relativity and quantum theory, to move up through its ruthless refereeing machinery.

The exceptional course of events in the intelligent-design case, in which complete books were published on their new findings, worked like an alarm. The biological community was prompted to try and explain the examples from evolutionary principles, and the irreducible complexity did not turn out to be quite as irreducible as the proponents had claimed. The controversy culminated in the Kitzmiller lawsuit filed with the U.S. District Court for the Middle of Pennsylvania in

2005, where eleven parents sued the Dover Area School District over the plan to teach ID in their science curriculum. The case led to a strongly worded verdict by Judge John E. Jones III, who produced a very thorough 139 page document substantiating his conclusion that teaching intelligent design in public school biology classes would violate the first amendment of the constitution of the US, because intelligent design is not science and cannot decouple itself from its creationist and thus religious antecedents.

The lesson that I have learned from this incident is that it is always dangerous to draw huge conclusions based on what we *don't* understand; it is preferable to draw modest ones from what we *do* understand. The reason why I mention the case here is that it took the public arena by surprise and could have done serious harm. The intelligent design movement, viewed by many as outright antiscientific, fiercely insisted that their view on biology had to become part of the standard curriculum in high schools, to counterbalance the implicitly "nihilistic, antireligious" content of the present science programs based on evolution. I can only en-

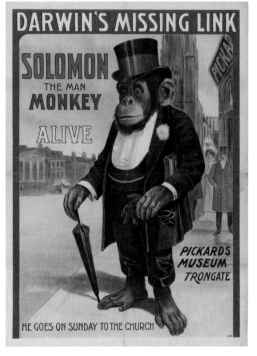

dorse the strongly worded reaction of many scientific organizations in the US and elsewhere, which called on parents and teachers not to approve "any lesson plans that require students to compare the well-tested science of evolution with the dubious hypothesis of intelligent design. Religious doctrine – in any guise – does not belong in science classrooms." Already in 2002 the very serious popular-science journal *Scientific American* published an unusual article entitled "15 Answers to Creationist Nonsense," that began by stating:

> *When Charles Darwin introduced the theory of evolution through natural selection 143 years ago, the scientists of the day argued over it fiercely, but the massing evidence from paleontology, genetics, zoology, molecular biology and other fields gradually established evolution's truth beyond reasonable doubt. Today that battle has been won everywhere – except in the public imagination.*

You might expect that after centuries of scientific progress, the bankruptcy of myth and superstition would need no further comment, but this is far from true. The respectable journal *Science* reported in August 2006 that the public support for the theory of evolution in the US was actually decreasing! In the period 1985-2005 the percentage of the adult population that said evolution theory was "true" dropped from 45 to 40 percent. The fraction that rejected Darwin's theory also decreased, from 48 to 39 percent, and the percentage of people who were "not sure" tripled from 7 to 21 percent. It seems that the Middle Ages have never really ended. But the real problem was also pinpointed in that same article:

> *When presented with a description of natural selection that omits the word evolution, 78% of adults agreed to a description of the evolution of plants and animals. But, 62% of adults in the same study believed that God created humans as whole persons without any evolutionary development. It appears that many of these adults have adopted a human exceptionalism perspective. Elements of this perspective can be seen in the way that many adults try to integrate modern genetics into their understanding of life. For example, only a third of American adults agree that more than half of human genes are identical to those of mice and only 38% of adults recognize that humans have more than half of their genes in common with chimpanzees. In other studies fewer than half*

of American adults can provide a minimal definition of DNA. Thus, it is not surprising that nearly half of the respondents in 2005 were not sure about the proportion of human genes that overlap with mice or chimpanzees.

We could proceed to a next round about the conflicting interests of religion and science in modern culture. For this, I refer to some interesting books like *The God Delusion* by Richard Dawkins and *Breaking the Spell: Religion as a Natural Phenomenon* by Daniel Dennett. Entering into that debate would take me too far a field, so let me close the subject with some balanced remarks on this issue by Jerry Coyne of the University of Chicago:

But the acceptance of evolution need not efface morality or purpose. Evolution is simply a theory about the process and patterns of life's diversification, not a grand philosophical scheme about the meaning of life. Philosophers have argued for years about whether ethics should have a basis in nature. There is certainly no logical connection between evolution and immorality. Nor is there a causal connection: in Europe, religion is far less pervasive than in America, and belief in evolution is more widespread, but somehow the continent remains civilized. Most religious scientists, laymen, and theologians have not found the acceptance of evolution to impede living an upright, meaningful life. And the idea that religion provides the sole foundation for meaning and morality also cannot be right: the world is full of skeptics, agnostics, and atheists who live good and meaningful lives.

There are many other dramatic instances in which responsible people and institutions adopt ludicrous mythical interpretations and measures, even though clear-cut rational solutions are available. One of these is the fatal rate at which the HIV virus (see page 53) has been spreading in Africa. Millions of people have become the victims of an immensely arrogant refusal to accept scientific evidence as a solid base for action. There may be a sociological explanation for this, and perhaps civilized people should respect other people's views. But if those people happen to be in power and they bear the responsibility for the lives of millions of other people, then respect is not the first word that comes to mind.

Wangari Maathai, vice-secretary for the environment in Kenya, won the Nobel Peace Prize because she devoted her life to laudable causes like organizing opposition against deforestation and the propagation of women's rights in Africa. I find it utterly painful to hear that she has been making the most horrendous statements about HIV and AIDS. Ms. Wangari Maathai has declared HIV to be a biological weapon deployed against the black race, specifically designed by certain Western scientists to gain control over Africa. For a long time President Mbeki of South Africa, where AIDS has been the number one death cause for a long time, too held absurd views on HIV, which made effective action against the disease basically impossible.

Also the Vatican has repeatedly argued that the use of condoms is one of the main causes of the rapid spreading of the virus and should therefore be avoided. This was put forward again recently by Pope Benedict XVI in a public address during his first visit to the African continent in March 2009. His claim that the AIDS problem "cannot be overcome with the distribution of condoms which, on the contrary, increase the problem," has given rise to strong reactions from health and political organizations. The influential medical journal *The Lancet* featured a strongly worded editorial calling the statement "outrageous and wildly inaccurate" and accusing the church father of "distorting scientific evidence to promote Catholic doctrine."

Should lives ever be sacrificed for some doctrine, just because rulers fail to recognize reduces the risk of infection by 90% according to the WHO.

Finally there was an atrocious, primitive myth being spread that copulating with a virgin could exorcize HIV and heal you. Predictably, this led and still leads to numerous young girls being raped and infected. A criminal appeal to irrationality in an attempt to cope with fear and despair; traumatic for the victims, and tragic because it was bound to fail.

Fighting fear

The stories I have presented in the previous section are but a few out of a large repertoire. In many ways they are variations on Benjamin Franklin's "pagan rod" or Steinberg's "Ninth avenue perspective," albeit on a rather devastating scale. We have seen that myth can be a powerful weapon in the public arena. It may be generated by despair, but we have also seen that fear can be exploited in order to consolidate questionable power structures. The stories also painfully show that myths can be extraordinarily persistent. This persistence in the public domain is fueled mainly by ignorance, as ignorance is the mother of fear. Nobel Prize winning economist Paul

truth and reason? At an earlier similar occasion, the World Health Organization reacted by issuing a declaration:

These incorrect statements about condoms and HIV are dangerous when we are facing a global pandemic which has already killed more than 20 million people, and currently affects at least 42 million.

Things may sometimes go wrong in using condoms, but the virus truly cannot move through them; consistent and correct usage

Krugman made the following observation in a column for *The New York Times*:

> *There are several reasons why fake research is so effective. One is that nonscientists sometimes find it hard to tell the difference between research and advocacy – if it's got numbers and charts in it, doesn't that make it science? Then there is also the dubious contribution of the press referred to as he-said-she-said journalism which gets in the way of conveying the correct image of science, just because the reader wants the story more juicy: You can imagine that if the president of the US would say that the Earth was flat, the headlines of news articles could read, "Opinions Differ on Shape of the Earth."*

The dramatically distorted public image of science might be exactly what motivates the numerous outreach efforts made by an increasing number of committed scientists. To wit, for me personally, the frequent distortion and denial of well-established scientific knowledge has been a main motivation to write this book. I like to think that these scientists manage to convey the key messages of real science and of scientific culture to an ever-larger audience distributed over many layers of society. And that is what is needed. New ideas tend to meet strong opposition, because preconceptions and prejudices are often extremely persistent. The defenders of the *status quo* are usually those who have vested interests in it. Fortunately, youth can

young lady posing behind a typewriter. The fact that such a book was written – and presumably sold as well – implies that there must have been a non-negligible fraction of girls that were not quite happy with the role model that was imposed on them. And fortunately, in the meantime things have changed, at least in some parts of the world. There we now find men who get scared from posters like the one on this page and think that we are entering an era in which women will take over the Fortune 500 and the world will be completely feminized. The issue of women rights and the implementation of them is also a cultural domain where the pursuit of truth has been suppressed in a long-lasting power struggle. This too is a case in which transcending the myth would benefit all, but up till now prejudice mostly prevailed.

Let us end this chapter with an encouraging epilogue. The debate between scientists and the league of "there is more between heaven and earth" adepts has deep roots that go back to antiquity. There are even some instances of such debates in the Middle Ages, when science was paralyzed by incontrovertible dogmas entirely fixed by the

make us aware of the poor rational justification of many of our views by revolting against them, sometimes in defiance, sometimes just by living a very different life.

The picture of the woman at the stove stood for the female role model expressing sheer happiness only 50 years ago. At that time I remember a book appearing with the title *If only I were a boy*, with on the cover a sad teachings of Aristotle, which were canonized by the Roman Catholic church. Indeed, centuries before Galileo Galilei defended his support for the Copernican model of the universe against the Inquisition in 1633, there had been remarkable insurgences challenging the religious doctrines of those days.

You may hear modern scientists refer to *Ockham's razor*, which refers back to the mortalist William of Ockham (1285-1347). He decreed that "*entia non sunt multiplicanda praeter neccesitatem*," meaning that "entities must not be multiplied beyond necessity." This principle of parsimony defines a sort of scientific minimalism: it advocates the importance of using the absolute shortest encoding for the description of phenomena. It is a plea for scientific clarity and efficiency: models that employ more variables and parameters than strictly necessary should be abolished, and that is pretty much what modern science amounts to.

Another remarkable character was Abelard (1079-1142), in fact a medieval precursor of the Renaissance. He was well known for his very unfortunate love affair with the much younger Heloise, which ended tragically.

When their child named Astrolabe was born, the loving parents were forced to separate. Abelard was castrated and both Abelard and Heloise withdrew in monasteries.

Now this Abelard in his *Sic et non* already postulated four golden rules for research and debate in the early twelfth century:

- *Use systematic doubt and question everything.*
- *Learn the difference between statements of rational proof and those merely of persuasion.*
- *Be precise in use of words, and expect precision of others.*
- *Watch for error, even in Holy Scripture.*

Abelard published a list of no less than 158 philosophical and theological questions on which the opinions in those days were clearly divided. With this dialectic approach he was a highly appreciated teacher, who can be said to have laid the foundations for modern western education. He was an important innovator in that he dared to challenge the scholastic tradition in which the teachings of the church were incontrovertible.

Much later, but still well before Bruno and Galilei, Pietro Pomponazzi (1462-1525) from Padua tried to eradicate black magic and superstition in his book *De Incantationibus* published in 1520:

It is possible to justify any experience by natural causes and natural causes only. There is no reason that could ever compel us to make any perception depend on demonic powers. There is no point in introducing supernatural agents. It is ridiculous as well as frivolous to abandon the evidence of natural reason and to search for things that are neither probable nor rational.

These words have not lost any of their meaning or relevance even today, and I would gladly nail them to the door of many a classroom. It goes to show that demystifying forces have been present as long as there have been people around who dared to think independently, even though they often had to pay a high price. If only we would learn from the past …

Basic questions

Many people associate science directly with technology. This, however, leaves an essential part out of the picture. There exists an intimate relationship between the two, but they are complementary in several essential ways. As I tried to convey in the first chapter, the first ingredient of science is wonder – that is, posing questions out of sheer curiosity. These questions may be very down to earth and concern things we can touch or at least see. But we also ask questions about very abstract notions. Have a look at the table on the following page: the questions that are listed in the left column are so basic that everybody is bound to run into them at some point in their life. Many people will already start asking them in their childhood. These questions are universal: they are of all times and have been asked in some form or another in almost all cultures, where they have given rise to a rich spectrum of religious, artistic and scientific expressions. They have always been conceived as important and form a basic part of our view of what a culture really is.

Good answers to these questions are not easy to give. Even if you know an approximate answer, it's hard to decide where you should start the explanation. You may find it surprising that no conclusive answer has been given to any of these questions. In spite of thousands of years of science, there is no absolute certainty about any of these matters. What we have harvested so far is rather a great variety of would-be answers demonstrating the great ingenuity, creative powers, sense of beauty and even humbleness which we as humans apparently possess. We have found very different ways to "deal" with these questions, diverging in their very nature.

What is...?	Myth	Science
matter	Alchemy	Chemistry Physics
space and time	Astrology	Astronomy Mathematics
the universe	Creation	Astrophysics Cosmology
life and death	Witch doctors	Biology Medical sciences
mind and soul society	Religions	Psychology Sociology Anthropology
	Mystical, esoteric, spiritual, belief	Realistic, materialistic, logical, critical, empirical

The list has a property that may have escaped your attention: from top to bottom the questions become harder and at the same time more vague, with an increasing spiritual component, you might say. On the other hand it is worth noting that the overall spiritual component of these questions (and answers) is decreasing over time.

Supply and demand

The topics listed in the table are quite generic. Nevertheless questions about them have plagued mankind for as long as we know. It is clear that there has always been a strong need for answers to these questions.

Although man had consciousness and the ability to think, he found himself in the uncomfortable situation that he could not (yet) understand the world around him, let alone predict or manipulate it. The desire to control one's physical environment is quite natural, based on the drive to establish stability and security, two important evolutionary parameters. The human condition has always been characterized by an existential gap: man was part of something larger that he could not understand. In early history he was completely subjected to Nature's arbitrary actions. The basic realities of life were governed by a kind of economic law even then: if there is a demand, it will generate a supply. Whether it was motivated by love or contempt for our fellow humans, a desire for purity of the soul, or just the pursuit of profit, as long as we have existed there has been a continuous flux of highly appealing "answers" to the questions listed. The existential gap was also a gap in the means-for-survival market. And so an impressive variety of tempting answers appeared on the market (and still does), mostly in the form of myths, (pseudo)science, superstition and religion. Human history shows a rich spectrum of magic spells, rituals, recipes and the like aimed at making life bearable for the ignorant and the deprived by providing hope and comfort.

At the mythical stage, each of the questions led to what I like to call *Holy Grail* phenomena – extremely attractive, but unattainable options. To start with the first question: we know that the problem of understanding matter gave rise to alchemy, a branch of investigation and knowledge in search of the "stone of wisdom," which would allow people to turn inferior metals like lead into gold. Amusingly, this search even gave rise to an early version of the "gold rush." It became such big business that in 296 AD the Roman emperor Diocletian decreed that gold making was forbidden and all books about the subject had to be burned! Incidentally, nowadays we would in principle be able to make gold out of other stuff, but it would be much more expensive than just buying gold.

The questions about the universe became linked with astrology. The orbital motion of the planets was known in detail and thus highly predictable. The magnificent idea was to somehow relate people's personal history to the predictable motions of the planets and their ascendants. That would allow people to extrapolate and predict their future. If true, this would enable them to eliminate fate, and as you know this is still believed and practiced on a large scale by a multitude of astrologers and their followers today.

Questions about space and time often mingled with questions about the universe. There is an abundance of "creation myths," which differ greatly in nature. The fact that they are in many ways contradictory did not really matter, because these myths were confined to a specific place and time. However, a common theme is the concept of some superhuman creator. In that sense these creation myths also took care of the difficult question of the meaning and purpose of life. Why are we here in the first place? In fact the call for a direct and simple answer was transferred to a higher being which existed somewhere, but with whom we could not directly communicate. Here we enter the subject of religion in its many guises, which is also the realm of mind and soul. A Supreme Being (or a collection of them) was somehow in control of reality. Often the essence of this being's teachings was expressed in a holy book which constituted the sole source of truth and had to be read and explained by prophets and priests. This marks the beginning of schools of theology and philosophy that tried to straighten out these highly spiritual, not very tangible matters. People were given the feeling that there was a purpose and that somebody high up was looking after them – and maybe at some final instance judging them. It provided hope and solace and helped them to cope with the misery and mystery of life on earth, for example through the prospect of a magnificent afterlife.

The perplexing question of life and death was the domain of witch doctors. Here the Grail was the search for an "elixir of life," which would allow people to acquire immortality. Basically it was the mythical response to a fear of illness and death. The secrets of how these stones and elixirs were to be found or brewed were kept by a spiritual elite of priests and wise men or women. The aim was to prevent crooks from running off with these awesome powers and selling them to random people who could bring about all kinds of disasters.

When we now read books and accounts about these endeavors – which manifested themselves in the West at least until the Renaissance – we can only be impressed by the immense creativity of those spiritual men and women, and at the same time be quite amused by the obscure mixture of reason and belief underlying it. Yet these practices hinted at solutions to very practical problems: finding the "stone of wisdom" certainly would free you from all your material concerns. Even today people still seek ways to achieve that, in equally mysterious practices. Many of us who do not really understand economics prefer to follow some business guru. And our fear of physical suffering and death would still be quenched by an "elixir of life."

Religion in all of its manifestations not only gave meaning to our existence, it also provided people with a set of rules telling them how to fulfill the task set out for them. It gave them a means to justify their existence by living an exemplary life, and furthermore informed them what such a life exactly amounted to.

Altogether we may say that myth was a social construct aiming to eliminate the role of arbitrariness and fate from the human condition, a condition characterized by our splendid abilities to survive, observe and think, combined with a total lack of understanding. The overall principle in the mythical era was that you had to believe what you were told, according to certain revelations. The truth existed, it was fixed once and forever, and was (partially) known to certain experts but could only be shared in a mysterious, esoteric, and implicit way. Myth provided mental comfort to some, but at the same time it led to suppression of thought for others. Deviating from the common belief could land you in very serious trouble, from imprisonment to burning at the stake.

Empirical science

> *My work, which I have done for a long time, was not pursued in order to gain the praise I now enjoy, but chiefly from a craving for knowledge, which I notice resides in me more than in most other men. And besides, whenever I found out anything remarkable, I have thought it my duty to put my discovery down on paper, so that all ingenious people might be informed thereof.*
>
> *Antoni van Leeuwenhoek*
> *Letter of June 12, 1716*

We now turn to the right-hand column of the table, which shows the – shall I say – *modern* account of how we deal with the basic topics listed. And of course this is a list of sciences. To discuss matter scientifically, typically we would consult a physicist or a chemist. To address space and time we turn to mathematics, physics, and astrophysics, and the story of the universe is likely to be written in the fields of physics and cosmology. Most of us nowadays prefer to go to the hospital instead of the witchdoctor. If herbs happen to possess excellent healing effects, biology and the medical sciences mostly allow us to understand why. Questions about our mental well-being, as an individual but also as a social collective, we try to address by studying psychology, sociology or anthropology – though there are still plenty of people who turn to alternative means for these issues.

The essential difference between the two columns lies in the methodology employed. In science, high values are placed on critical analysis and logical reasoning, and ever since the era of "Enlightenment" in the 17th century we cling to experimental verification. Since those days, the crucial role of observation as the final arbiter in the intellectual debate in order to achieve progress, has become more and more evident. Our search for the truth became an effort in which big questions were broken down into smaller ones, which were to be resolved in direct dialogue with nature itself, rather than with a

Supreme Being through a ritual intervention of some initiated person.[12]

The table may overemphasize the differences, but – and this, to me, is a very nice aspect – it also shows clearly that the roots of the modern sciences reside in the mythical domain. Alchemy was the cradle of chemistry, et cetera. This is a sobering thought for a hard-core practicing scientist like myself. Nevertheless, in an essential way the approaches are antipodal. The mythical approach is top-down: an incontrovertible truth is handed down from some highest instance and has to be believed *a priori*. The scientific approach, in contrast, is bottom-up: one starts by asking questions and exploiting our humble human ability to observe and think, and tries to slowly acquire knowledge by reading the "Book of Nature" over and over again. Scientists continuously make a tremendous effort to represent the facts in a logically consistent framework, which eventually yields an understanding that we call a theory of the world. This process of ever-growing insight implies a dynamical notion of truth.

Does this mean an end to the mythical way of thinking? Not necessarily. Niels Bohr, nuclear physicist and winner of the Nobel Prize in 1922, once defended a horseshoe hanging over his door by claiming: "They say that it works, even if you don't believe in it."

I will now discuss how the scientific process works and how it has brought us such remarkable progress in our understanding of the world around us, even though its truth is never fixed.

The knowledge machine

Let me take you on a small excursion into meta-science. I want to present a simple picture of how science works, often referred to as the *scientific method*. I used the term "double helix," an image borrowed from our modern knowledge of the structure of DNA,

[12] It is interesting to note that also in the Greek culture this emphasis on experiment developed after Plato died. It was in the Lyceum of Athens rather than the Academy, that natural philosophers like Aristotle, Archimede, Theophrastus and Strato were proponents of this type of investigation. It was halted after the Aristotelean world view had been canonized by the Catholic church and no further discussions on the matter were allowed, let alone stimulated.

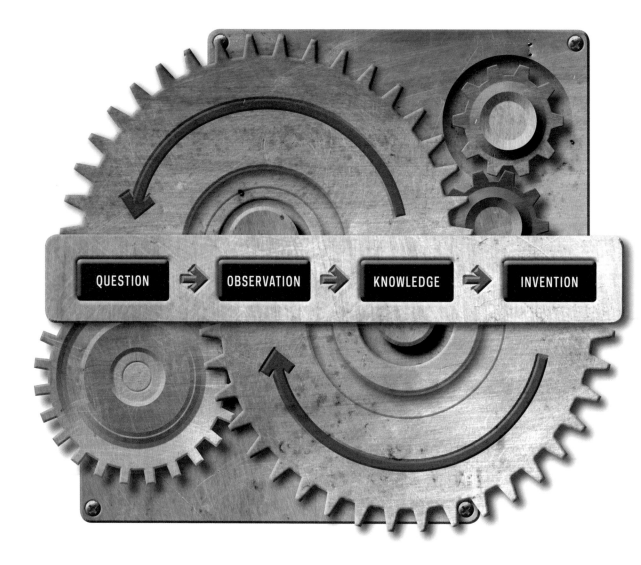

QUESTION → OBSERVATION → KNOWLEDGE → INVENTION

because of the way in which science and technology play complementary roles in the process of knowledge generation.

In the diagram on the previous page you can see two cycles, one at the top and one at the bottom. In the middle there is the sequence "question, observation, knowledge, invention." The top cycle is the knowledge cycle, which could be described by Francis Bacon's statement "Wonder is the seed of knowledge." The bottom cycle is the technology cycle, which can be described as "Knowledge is the seed of technology." The top cycle is "curiosity driven," the bottom one is driven by the possible applications. In an older language one would describe them as "pure science" versus "applied science." But I am getting ahead of myself. First I would like to go through this picture step by step.

Let us start on the left of the central bar, with the questions that reflect our wonder and curiosity. If we begin investigating the subject of our wonder, our subsequent observations may lead to knowledge. "Measuring is knowing," used to be the device of Heike Kamerlingh Onnes, who in 1911 much to his surprise discovered that liquid helium has the miraculous property of becoming superconducting at very low temperatures. Indeed observation leads to knowledge, but the top arrow pointing backwards is equally important. It symbolizes the feedback in the process: from new knowledge inevitably new questions arise. In fact this is not so much a closed loop as a spiral – our picture is only a two-dimensional rendering of a three-dimensional curve: the loop rises up from the surface of the page in the time direction, so to say, forming a spiraling corkscrew shape or a helix.

Indeed this cyclic property is obvious: once we discover that the earth is not flat, but a sphere, we are bound to ask what happens to people on the underside (the *antipodes* mentioned before). Do they exist, and if so, why is it that they don't fall off? Is there perhaps some force that attracts them to the earth? It is clear that the top cycle does not describe all of science. That single helix would stagnate at some point, because it would become impossible to make the required observations to progress further. And that is why we have to add the other, connected helix. I will illustrate this using the wonderful drawing by

M.C. Escher, showing a droplet of dew on a leaf. The picture has three layers of reality: there is the leaf with the dewdrop, and there is the world reflected in the droplet's surface, showing the windows in the room in which the plant is apparently growing – the space of the observer. Finally there is the part of the surface of the leaf that can be seen under the droplet, showing the veins of the leaf in great detail as the water magnifies what is below it. How wonderful! The curious mind has observed something very special indeed.

Once he has seen the magnification effect due to the droplet, the observer might think of something else. In the figure on this page you can see that glass too is able to deflect light rays in all kinds of ways: the square pattern of the tablecloth becomes strongly deformed if you look through the glass vase.

The only step needed to become a great inventor is to combine the two observations into a new question: might we be able to create magnifying droplets of glass, and perhaps even make them better so they give a clear image and a large magnification? That would almost make you the inventor of the lens, an absolute phenomenal discovery! Think of it ... The lens made it possible to develop a telescope, which provided a new view of the heavens. But the lens also allowed the construction of the first magnifying glasses and microscopes; it opened the door to the world of tiny entities, the microcosm. That is not all; the lens allowed us to make spectacles, with the wonderful result that for many people in this world their personal horizon got extended and their quality of life was improved enormously. Once the relevance of lenses became clear, it was of direct interest to study the laws governing the refraction of light in great detail. Thus the lens is a dazzling example of the interplay between science and technology, leading to – as we like

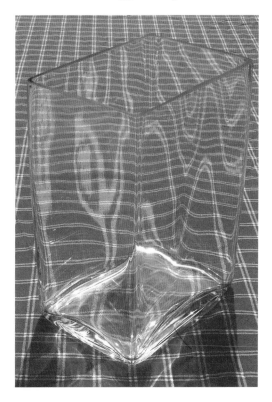

to call it nowadays – "societal validation."

Now you can see that the process described is in fact a second spiral: observation leads to knowledge, which leads to an invention, allowing for new and better observation. And hopefully you also understand why the two cycles are intertwined, forming a *double helix*: the instruments obtained allow us to increase the range of our observations. The lens allowed researchers to see what nobody had seen before, such as moons of Jupiter and bacteria. These new observations led to new answers, which of course gave rise to new questions.[13]

Looking at this process from a distance, we can see that this process of systematically – and objectively – elevating the level of our observations is an important aspect of the enormous progress science has made ever since the Renaissance. The story of optical telescopes is not unique. They allow us to detect information carried by the light that is emitted by distant sources, basically just like when we see a flower or a bicycle. Now, since the end of the nineteenth century, we know that light is simply one particular form of electromagnetic radiation. There

[13] My tale does not do justice to the historical order of events. It presents a number of innovations in a single stroke, while in reality it took many centuries to get from a magnifying glass to the first telescopes and microscopes. I like the idea that the first person to consciously see the magnifying effect of a droplet of water made a crucial observation and deserves a lot of credit, but who would that have been? Maybe Eve – after all, she was clearly driven by curiosity. It is documented that the Greek philosopher and historian Seneca read all that was written in his time through a glass bowl filled with water. That story clearly shows that already in antiquity the magnifying properties of a body of fluids were known. The first magnifying glasses appeared in the early Middle Ages, mainly for the upper classes, as they could read and reached a respectable age. The invention of early microscopes and telescopes is often attributed to Hans Lippershey, Zacharias Janssen and Jacob Metius.

Galileo soon heard about the "Dutch invention," improved it significantly and used it to make his astronomical discoveries, like the moons of Jupiter.

The most famous early microscopist was probably the fabric trader from Delft, Antoni van Leeuwenhoek, who produced microscopes enlarging up to 480 times! He is considered to be the founder of microbiology, having made early reports on a large variety of microorganisms, from sperm to bacteria.

are other kinds of electromagnetic radiation that may carry additional information about those distant objects: infrared and ultraviolet radiation, radio waves and microwaves. Once detection devices were developed for these types of information carriers, researchers obtained totally different views on our universe, which offered important new clues about its content and history.

Scientists also learned that the earth atmosphere is a severely limiting factor to their observations of the sky, because it absorbs and scatters radiation. Therefore modern observational astronomy has moved its telescopes into outer space. We have all seen the incredible pictures the Hubble space telescope has sent us from deep space. If you check the websites of the space agencies, you will find that there are over fifty of such telescopes circling the earth and gathering data. Nowadays we even go beyond electromagnetic radiation – and why not? Every form of energy emitted by distant sources that we can intercept may provide us with unexpected information. So neutrino telescopes are constructed to detect the elusive massless and extremely hard-to-detect particles from outer space called neutrinos, and gravitational wave detectors will be used to spot very violent events in the universe such as the formation of black holes and the clashing of galaxies.

The account I have just given about our observing the universe at large can also be given about our investigation of the microcosm. There we moved from optical microscopes via the domain of electron microscopes to scanning tunneling microscopes. We even created accelerators that now allow us to see what is going on at extremely small scales down to about 10^{-20} meters. Also in medical science there have been spectacular revolutions thanks to the development of more and more refined diagnostic tools, allowing doctors to probe the structure and function of the human body on all scales. An intensive-care unit now somewhat resembles an advanced physics lab. Any modern hospital proudly possesses an ever-greater variety of scanning devices, such as ultrasound, CT, PET, X-ray and MRI. I present these examples just to show how far we have come: the double helix of science and technology has made many turns indeed over the last five hundred years.

I have used the lens and its descendents to clarify the back arrow at the bottom of the technology cycle. But this loop can also stand for entirely different "devices." Think for example of the discovery of the structure of DNA by Watson, Crick and Franklin in 1953. This event was like the opening of a great door to the chemistry of life, and has led to the possibility of studying life on a molecular level in a very systematic way. Nature provides us with a magnificent tool kit in the form of enzymes, allowing us to cut and paste the molecules of life. Using it, researchers are slowly decoding the molecular messages that we all carry around in billions. Here too, the technological output is spectacular, from genetic engineering to diagnostics and forensic applications. These are indeed turning points in our ability to manipulate nature, which have given a tremendous swing to our double helix of science and technology.

Yet another example of the technological feedback loop is the invention and rapid development of computers, which allow us to analyze data on an unprecedented scale. The advent of the computer has profoundly enlarged the scope of all sciences. From the classical synergy between experiment and theory, we have moved to a situation in which the computer has entered as a powerful third partner; a perfect *ménage a trois*, I would say. Computer scientists and staticians have developed very powerful techniques to do all kinds of data mining, using machine learning techniques or neural networks to search for regularities in large bodies of data. These regularities may provide hints on relations that allow for simple ways to model the data. Computers not only allow for massive manipulation of data, but also, through large-scale simulation studies, let us investigate theoretical models which cannot be handled by standard mathematics. Not only conventional science and engineering, but also the sciences of complexity like economics, psychology and sociology have profited enormously from these developments. Quantitative research on an unprecedented scale is changing many of the social sciences into data-driven "hard sciences."

I would like to mention one other aspect of the double helix, an aspect I particularly like.[14] It clearly is a machine that generates knowledge and technology, and it does so bottom-up. And once that machine is running, it becomes irrelevant how it got started: the left cycle drives the right cycle and vice versa. It is a rather autonomous machine that is hard to stop. Nobody ordered the stars, or DNA, or the atom; it was our insatiable quest for knowledge, that – via the long and winding road of the double helix – brought us the discoveries that have irreversibly changed our lives and our perspective on the cosmos. We could say that the far left of the illustration shows the very theoretical, maybe even philosophical end of research, while the far right might be the corner of the product improvers. But philosophy dries up if it is not fed by new insights from reality, and product improvement needs fundamental new input if it is to be more than a dead-end street. New ideas are indispensable to keep playing an active part in a world where the future is in the hands of the most innovative entrepreneurs: it doesn't pay to try to improve the electric calculator if somebody else is producing a pocket-size semiconductor device that can do a thousand things more, a thousand times faster without producing any noise. To stay on top of things, you need to join the double helix.

What science cannot do

In conclusion, I am unwilling here to say anything very specific of the progress which I expect to make for the future in the sciences, or to bind myself to the public by any promise which I am not certain of being able to fulfill; but this only will I say, that I have resolved to devote what time I may still have to live to no other occupa-

14 Not surprisingly, as the development of science and technology is high on the agenda of national and international governing institutions, many books have been written describing similar models for growth and progress. Some of these are quite detailed and refined, others can be seen as dressed-up versions of a single idea. To my knowledge, Hendrik Casimir, manager of the physics laboratory of Philips, was the first to describe the "spiral of science and technology," in a talk in 1952. The extended versions typically include other players in the science-technology game, such as private enterprises or government. When their in- and output are included, this leads to triple helices and other structures reminding of the dazzling complexity seen in molecular biology.

tion than that of endeavoring to acquire some knowledge of Nature.

Descartes

Up till now, I have emphasized what great things science can achieve. But the nature of the helix immediately shows that it is a process: the discoveries follow each other in a rather predetermined order. It is hard to discover quarks before you have discovered the atom, and it is equally hard to understand galaxies without first knowing what stars are, to build computers without an understanding of the quantum mechanics of semiconductors, and so on.

So even though society wants something very much, science is often not able to provide it. It's not like a shoe store or a car factory. If some agency declares that cancer has to be eradicated in five years' time, or that we need fusion to work right now, the project is bound to fail. It is simpy too hard to skip nature's fine print and still jump to truthful conclusions. Many branches of science do have their own Holy Grail: artificial intelligence, the physical "theory of everything," total control over diseases, *et cetera*. But only in the very long run can we see clearly that enormous progress towards those laudable goals has been made. Science is a historical process, coevolving with our species, and it is not something that is easy to steer, although government agencies might claim the opposite. Persons or institutions in power may stimulate or delay certain developments, which may or may not lead to tangible results, but those interventions are more like perturbations on the historical lines of how nature is stripped bare layer after layer. To understand the path of science as a whole, it pays to think in terms of fifty or even a hundred years, which is clearly beyond the life expectancy of governmental agencies.

Now you understand why I chose to open this chapter with the picture of a long spiral staircase. Yes, indeed it is like a stretched DNA molecule and maybe it also has three billion steps, if it is finite at all. It may have struck you, that looking at it in another way, it is just an eye, a human eye that is patiently observing you, the reader.

Science is a unique human endeavor. To succeed, it takes a strong motivation and perseverance, apart from curiosity, integrity and ingenuity – from society as a whole but also from the individual scientist. I would like to conclude this chapter with a quote from Max Perutz, the founder of molecular biology, who came to Cambridge from Austria as a young postdoc and received a Nobel Prize in 1965 for the discovery of the structure of hemoglobin. His description of the very early stages of molecular biology is impressive and moving, but in particular bears witness to the perseverance needed to be a great scientist:

Discovering the structure [of hemoglobin] was wonderful. You must imagine the time when proteins were black boxes. Nobody knew what they looked like. There I was, having worked on this vital problem for twenty-two years, trying to find out what this molecule looked like, and eventually how it worked. When the result emerged from the computer one night and we suddenly saw it, it was like reaching the top of a difficult mountain after a hard climb and falling in love at the same time. It was an incredible feeling to see this molecule for the first time and to realize that my work had not been in vain: because during those long years I feared that I was wasting my life on a problem that would never be solved.

Max Perutz

But in what modes that conflux of first-stuff
Did found the multitudinous universe
Of earth, and sky, and the unfathomed deeps
Of ocean, and courses of the sun and moon,
I'll now in order tell. For of a truth
Neither by counsel did the primal germs
'Stablish themselves, as by keen act of mind,
Each in its proper place; nor did they make,
Forsooth, a compact how each germ should move;
But, lo, because primordials of things,
Many in many modes, astir by blows
From immemorial aeons, in motion too
By their own weights, have evermore been wont
To be so borne along and in all modes
To meet together and to try all sorts
Which, by combining one with other, they
Are powerful to create: because of this
It comes to pass that those primordials,
Diffused far and wide through mighty aeons,
The while they unions try, and motions too,
Of every kind, meet at the last amain,
And so become oft the commencements fit
Of mighty things – earth, sea, and sky, and race
Of living creatures.

Lucretius, ± 50 BC
De rerum natura (On the nature of
things)
(Translation: William Ellery Leonard)

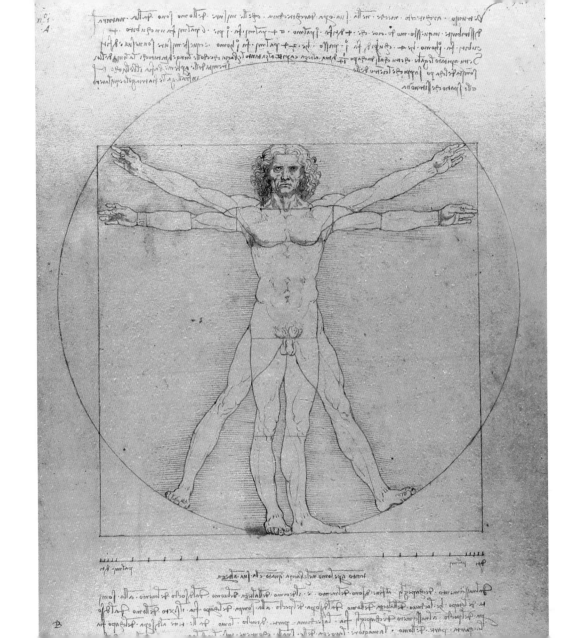

The Ouroboros of science

On page 83 you see a sketch of the mythical monster called the Ouroboros as it appeared on a 1975 preprint of an article by the American physicist and Nobel laureate Sheldon Glashow, then at Harvard University. In an imaginative way he combined the figure of the snake that swallows its own tail with a map of the natural phenomena we know about. The Ouroboros is an old Greek image that has been used by the Gnostics and also by alchemists to symbolize the unity of nature. The basic message is: To know and fully understand one thing is to know everything. The macrocosm has its mirror image in the microcosm. In Plotinus' *Enneads* (Fifth Ennead, 8th treatise) one finds the following somewhat elliptic description: "And each of them contains all within itself, and at the same time sees all in every other, so that everywhere there is all, and all is all and each is all, and infinite the glory."

The Ouroboros is the appropriate symbol for what I want to discuss in this chapter: the fundamental unity and connectedness of the natural sciences. The symbol gently reminds us of the mythical roots of the natural sciences. Also there are many echoes of it to be found in the course of history, such as in William Blake's little gem:

To see the world in a grain of sand
And a heaven in a wild flower
Hold Infinity in the palm of your hand
And Eternity in an hour.

In a more contemporary context, you may think of that awesome DNA molecule carrying all the genetic information that is required for the reproduction of the whole organism. Its information content is comparable to that of a large bookcase. This information is present in every cell of your body, so you can imagine your body as carrying around between 10 and 100 trillion identical bookcases. As everybody is carrying a hundred trillion identical libraries around, this can hardly be called efficient. It is an extravagant form of redundancy, quite different from having a spare tire in your car. However, when you think about how much information is piled up in such a tiny volume, the conclusion must be that at the same time an extreme form of efficiency is achieved. In this respect there are still quite a few lessons that we can learn from nature. On a symbolic level DNA exemplifies the Ouroboros idea that the whole is mirrored in the smallest parts.

On the following pages, I will use this Ouroboros as a template to highlight many different aspects of the natural sciences. We will uncover layer after layer of what the sciences are about, and I will point out ways to think about them in their diversity as well as in their unity. The image covers all basic phenomena that have been discovered at vastly different scales, and it depicts the sciences that have developed along with those discoveries. Looking at it from a distance, we will then be able to identify the truly essential turning points in scientific thinking about nature. We will also see how these are connected; this becomes particularly clear if we look at the Ouroboros in an overall time perspective. The Ouroboros of science offers a rich and unified perspective on the whole of nature, together with our knowledge and understanding of it.

The figure on the facing page summarizes the cyclical processes of the scientific enterprise which I discussed in Chapter II, but from a slightly different perspective. In the background you can see the circle of science with its many icons, which we will investigate from a variety of angles in this chapter.

The empirical scientific method is crucial to science. Besides experimental tools to make observations, it extensively uses the tools of mathematics, statistics and informatics. These together with a priori knowledge from theory form the input, and as before, knowledge and technology form the output. This basic methodological toolkit underlies the picture as a whole; each of the scales on the circle can only be explored because we have learned to implement our incessant quest for knowledge within a single procedural framework.

Now take a look at the icons on the circle. At the bottom in the center we find ourselves. Moving up to the left you encounter ever larger distances, covering the macrocosm, and on the right-hand side there is the microcosm. Loosely speaking, when you follow the circle from the human figures to the left as well as to the right, you see the historical path of scientific discovery. This process of slowly climbing both sides of the circle took many centuries, even millennia.

The picture raises some interesting questions. Why is the Ouroboros depicted as a circle, and not as a line, with ever-larger scales on the far left and ever-smaller distances on the far right? There is a deep reason for this, which will be revealed in due course – it is one of several deeper aspects that will be addressed towards the end of the chapter.

But let us first move on to the next overlay of our circle, where I will discuss the separate icons along the circle of nature. They stand for the various structures that make up our universe – the macrocosm on the left, and the microcosm on the right.

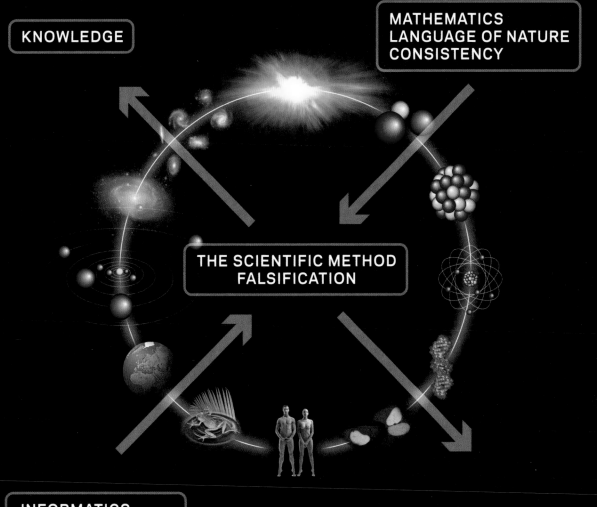

KNOWLEDGE

MATHEMATICS
LANGUAGE OF NATURE
CONSISTENCY

THE SCIENTIFIC METHOD
FALSIFICATION

INFORMATICS
SIMULATION
DATA PROCESSING

TECHNOLOGY

Outward bound: The macrocosm

What do we see when we move up the scale? Which structures did investigators encounter on their way out? This is an interesting story all by itself. Once upon a time people started out slowly, modestly exploring their immediate surroundings, by foot or maybe on horseback. Perhaps they were driven by curiosity, but more likely they were nomads searching for food and fertile grounds.

In antiquity the insight emerged that the earth was not flat, but a sphere, presumed to be perfectly at rest in the center of the universe. This is the *geocentric* universe, proposed by Claudius Ptolemaeus in his astronomical treatise *Almagest* in about 140 AD. It came as quite a shock when Nicolaus Copernicus – in his masterpiece *De Revolutionibus Orbium Celestium*, which appeared in 1543 – showed that the earth was not located at the center of the cosmos, but rather moved in an orbit around the sun like all the other planets.[15] Man had displaced himself and was no longer playing on center court. This idea of a *heliocentric* universe elicited massive resistance. Galileo Galilei took a firm stand in favor of the Copernican view in his *Dialogues concerning the two chief world systems* of 1632. He had to explain his views to the inquisition in Rome, and finally had to abjure his views. But in the margin of his copy of the *Dialogues* he scribbled:

Take note, theologians, that in your desire to make matters of faith out of propositions relating to the fixity of sun and earth, you run the risk of eventually having to

[15] It is interesting to note that even in early Greek times, Heraclides Ponticus (± 350 BC) already argued that the earth rotated around its axis in 24 hours and that the planets revolved around the sun. He suggested that the sun was in the center of the cosmos. A century later this was also put forward by Aristarch of Samos. Presumably the invincible status of Aristotle and Ptolemy prevented the idea from receiving the attention it deserved.

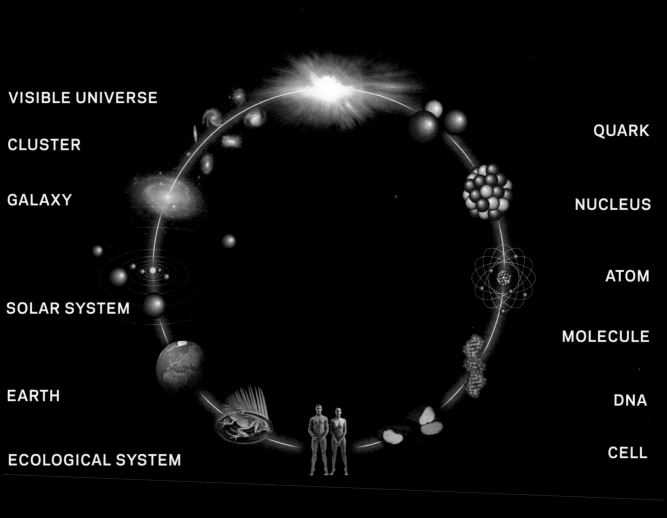

VISIBLE UNIVERSE

CLUSTER

GALAXY

SOLAR SYSTEM

EARTH

ECOLOGICAL SYSTEM

QUARK

NUCLEUS

ATOM

MOLECULE

DNA

CELL

HUMAN

condemn as heretics those who would declare the earth to stand still and the sun to change position - eventually, I say, at such a time as it might be physically or logically proved that the earth moves and the sun stands still.

Once this giant step was taken, the sun and all the planets orbiting around it could be studied as the *solar system.* It took another three centuries before people learned to appreciate that the sun was a star, just like the countless others they saw in the night sky. This realization cast great doubt on the notion of the sun as the unique star at the center of the universe.

In the 20th century scientists learned that at very large scales stars and their remnants were forming huge collective structures called *galaxies,* like our Milky Way, which contain up to 100 billion stars each. Then, studying even larger scales, they learned that galaxies form *clusters* too, which in turn form superclusters. Those clusters appeared to live on certain gigantic filaments and walls which form the boundary of incredibly large voids ...

And then we ran into the boundaries of the observable universe. Our universe may be much larger than the observable universe, but light from those remote regions has not had enough time to reach us yet! It was concluded from observations of our visible part of the universe that the universe as a whole must be very homogeneous: viewed on enormous scales it should be the same everywhere. Recently there have been speculations that actually this is not the case, and that we live in a *multiverse* with very many brother and sister universes. These could be vastly different physically, with different content, different laws and different histories.

Inward bound: The microcosm

On the other side of the circle you can see what is basically the hierarchical structure of matter – and as far as we know of *all* matter in our universe. There are two main parts to this side of the Ouroboros: the structure of living organisms, and the basic structure of matter. These two parts meet in the structure of large biological molecules such as DNA and proteins. At the bottom right of the circle you see the universal building block of all life, the *cell* – basically a volume of fluids enveloped by a membrane, containing the genetic material and working chemicals for the cell's purpose. *Prokaryotes*, like bacteria, consist of only one cell without a nucleus. The *eucaryotic* cells, out of which bigger life forms are built, have a *nucleus*, a region separated by a membrane from the rest of the cell, in which many other subcellular structures (organelles) are to be found. Examples are the *mitochondria*, producing the chemical energy for the processes taking place in the cell, and *ribosomes*, which help making proteins according to the instructions read from the DNA by the intermediate RNA molecules. In the nucleus we also find the DNA *molecule* which carries the genetic information, the *genes*. The cell and its contents, including complicated protein networks that are active within the cell, are studied in biochemistry, molecular biology and cell biology, together forming a field called system biology.

We are now moving up the right-hand side of the circle. In the question of what matter consists of, we have come a long way from the Greek belief that there are four basic elements: water, earth, air and fire.[16] The study of alchemy slowly evolved into modern

[16] This system of the four elements persisted for a very long time after Aristotle (384-322 BC), who proposed it. Interestingly, his direct successor Theophrastus (373-287 BC) already severely criticized the system, in particular the view that fire was an element. His clever argument was based on the simple observation that fire had no independent existence and durability. It had to be fed with other matter, and therefore could not be truly fundamental.

Aristotle's dominant views also eclipsed another fundamental idea, already put forward by Democritus of Abdera (460-370 BC). Democritus argued that everything had to be made up out of indivisible building blocks, which he called atoms, a rather stunning hypothesis at such an early stage of human thought. It would take another twenty centuries before this brilliant idea surfaced again.

chemistry around 1650. It remained difficult to understand the structure of matter, because it was hard to make direct observations and researchers had to think of clever experiments to make progress. For example, the French chemist Antoine Lavoisier (1743-1794) in his *Des Substances Simples* still wrote of *calorique*, meaning "heat," as an elementary form of matter, while we now know that heat is related to the kinetic energy of the moving atoms. *Lumière* or "light" also appeared on his list – and interestingly the idea of light as a substance is a view we share with him today.

In the intellectual history of chemistry, the work of John Dalton plays an important role. He proposed the term *atoms* again in 1803 and introduced the notion of atomic weights. Dalton understood that atoms could form molecules and that gasses like air were a mixture of different types of molecules. A great breakthrough was the periodic table, put forward by the Russian scientist Dmitri Ivanovich Mendeleyev (1834-1907). He organized all the known chemical elements in a grand scheme called the *periodic table of the elements*, managing to put all elements with similar chemical properties in columns of increasing mass. He could almost fill the whole chart, but there were a few open spots. For these he made firm predictions, giving the approximate mass and describing the chemical properties. Making predictions which are subsequently confirmed is the dream of every scientist.

Given that the elements fit so well into a highly ordered scheme, it became clear that there had to be a hidden principle or architecture explaining this systematic order. The concept of the *atom* finally fell into place, and it became an indispensable ingredient, both for explaining the properties of gasses and liquids, and for understanding the whole of chemistry.

The next step into the microcosm was the discovery by Ernest Rutherford in 1911 that the atom was in fact a composite object, and therefore not truly fundamental. It consisted of a positively charged *nucleus* around which negatively charged *electrons* were orbiting. The explanation of how such an atom could exist and remain stable was first given by Niels Bohr, whose atomic model constitutes one of the first great success stories of quantum mechanics.

Soon after the structure of the atom was unraveled, experiments showed that the nucleus itself was composite. The nuclei of all elements were built up from positively charged *protons* and so-called *neutrons*, which are electrically neutral. This amounted to a tremendous achievement. Why? Because the number of fundamental building blocks of matter was reduced from roughly a hundred in the periodic table of Mendeleyev to only three: the proton, neutron and electron. Reductionism never looked in better shape than in those days.

To study the properties of protons and neutrons and the forces acting between them, accelerators were built to speed up the particles and then smash them into each other. There were strong reasons to suspect that the protons and neutrons were also composite objects and would break up into pieces. But something strange happened: after smashing protons into each other with enormous energies, no smaller particles came out that could be interpreted as building blocks. Instead, many new species of nuclear particles appeared. It looked like we had arrived at the end of the story, where it was not possible to divide matter any further into smaller parts. Einstein's celebrated equation $E = mc^2$, expressing the equivalence of mass and energy, seemed to explain why. You see, the principle of accelerator physics is basically that if we give two particles an enormous kinetic energy, and then have them collide, a huge amount of energy is released, which is then converted into the mass of other particles that can be created abundantly in such collisions. So if the energy is higher, we gain access to heavier particle species.

Dozens of new unstable nuclear particles were discovered, all akin to the proton and neutron. An important new type of particle was the *pion*, a relatively light particle held to be responsible for the force which kept the nuclear particles together. We also discovered a big brother of the electron (about 200 times heavier), denoted as the *muon*, and a mysterious particle called the *neutrino*, which appeared to have neither mass nor charge, but nevertheless could carry energy and momentum. Of this phantom particle billions fly through your body per second; it interacts so weakly with other matter that it can traverse about a thousand kilometers of lead before being absorbed. And there were more new particles, with beautiful Greek fraternity-like names such as *rho*, *eta*, *phi*, *sigma*, *delta* …

It appeared that Einstein's formula somehow kept us from looking deeper inside matter: the more energy went into a collision, the more particles came out, but the proton would not break apart. We were confronted with a zoo of fundamental particles that did not seem to serve any purpose. These newcomers were not part of ordinary matter, because they would decay almost instantly into the known particles. When the muon was discovered, the famous nuclear physicist Isidor Isaac Rabi is reported to have exclaimed: "Who ordered that!" as if somebody had brought in a pizza with raw herring and peanut butter. Somebody had to come along to create order in the subatomic zoo.

Around 1963, Murray Gell-Mann and George

Zweig independently proposed that all nuclear particles could be understood as being composed of a very simple set of constituent particles: *quarks*. There were only three of them and they were called *up, down* and *strange*. Much later other types were discovered denoted as *charm, bottom* and *top*. The proton was composed of two ups and one down, while the neutron consisted of two down and one up. But nobody had ever detected these elusive quarks as free particles. "To be or not to be?", that was the question. If they did fly around freely, it would not be hard to detect them, because the up and down quarks had fractional charges of 2/3 and −1/3 the electron charge. The first hints that quarks do indeed exist came from subtle experiments in which electrons collided with protons, which revealed that the charge of the protons was indeed not localized in one spot, but instead on smaller entities, which could be identified with the conjectured quarks of Gell-Mann and Zweig.

The new enigma became why these quarks never come out of the proton. This puzzle is referred to as the problem of quark confinement, which is still unsolved and appears on the list of one million dollar millennium questions in mathematics, put forward by the Clay Institute in the year 2000. Even so, there is a theory that has already been successfully tested in computer simulations, but we would of course like to understand the confinement mechanism from first principles.

And that's where we are today. A new accelerator usually means breaking new grounds, and the physics community is particularly interested in the observations of what is at present the largest machine in the world, the Large Hadron Collider at the CERN laboratory in Geneva. Again we are exploring the unknown, and we may be served with an enlightening clue, or new mysteries may be revealed.

At the bottom of the circle we see a man and a woman, and their typical size is between one and two meters. We might say that the meter is our "order of magnitude" – which should not be a surprise, seeing that the unit is a human invention. But is man the measure of everything? If you think of all the structures we have been discussing, the answer is certainly no. To learn this, man had to venture into the world that surrounded him. On the left of the circle you can see the outside world, the macrocosm, where the scales grow ever larger, and on the right-hand side the inner world, the microcosm, with ever-smaller scales. How did scientists manage to measure objects of such very different sizes? That is a story of "inching along" with great ingenuity.

Distances in our solar system and nearby in our galaxy are measured by exploiting an elementary fact from plane geometry. If you know the length of the base of a triangle and you know the angles the base makes with the two other sides, you can calculate the height of that triangle. If your measurements are accurate, the height of the triangle can be much larger than the base you start out with. And the base can also be quite large: one end can be the location of the earth at midsummer and the other end can be where it is at midwinter, so that the base of the triangle can equal the diameter of the earth's orbit around the sun.

Another trick to determine very large distances is to use the apparent intensity of the light emitted by a special type of variable star called a Cepheid. These stars can easily be identified and on average emit the same amount of light. In that sense, they are like a standard 60 Watt light bulb. You probably know from everyday experience that the observed intensity of a lamp decreases as your distance to the lamp increases. Indeed, at a distance R, the light of the bulb has to spread over a sphere with radius R surrounding the bulb, and that surface increases proportion-

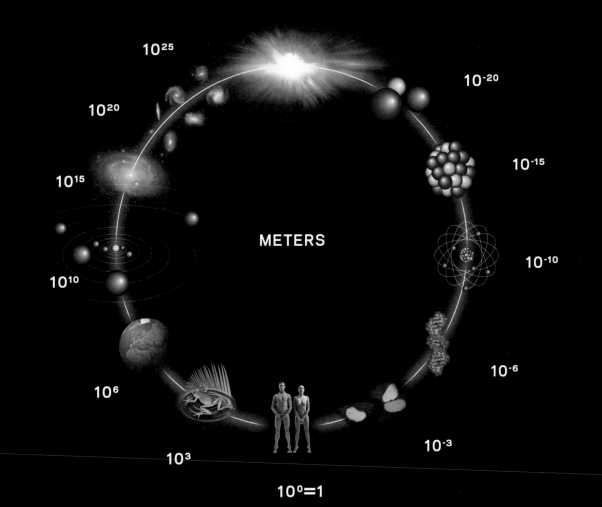

10^{25}

10^{-20}

10^{20}

10^{-15}

10^{15}

METERS

10^{-10}

10^{10}

10^{-6}

10^{6}

10^{-3}

10^{3}

$10^{0}=1$

ally to R squared. So if, for example, you move 3 times as far away, the light intensity drops by a factor of 9. This means that if you know the original intensity, you can infer your distance from the source by just measuring the light intensity where you are. Applying this trick to the distant Cepheids allows us to extend our distance measurements even to remote galaxies.

The largest distances in the universe are obtained by measuring the "red shift" of radiation. We observe a shift towards the red in the color spectrum of light coming from those very far objects, because they are moving away from us as the universe expands. The distance can then be determined using Hubble's law, which gives a very simple relation between the red shift and the distance.

The numbers around the circle quickly become enormous, though the notation used hides that. Once you understand that 10^2 denotes one hundred and 10^9 is a billion, you will realize that 10^{25} is a 1 followed by 25 zeros! I don't want to over-impress you, but the universe is Big indeed. Yet the notion of a large number is rather relative: for example, 10^{25} is also approximately the number of air molecules in a big birthday balloon. Still, 10^{25} grains of salt would be enough to cover the surface of the earth with a layer of salt of about 2 cm.

The number of air molecules in a balloon can be so big, because the numbers we encounter on the right-hand side of the Ouroboros are extremely small. When we go down in scale, we work with negative powers of ten: 10^{-1} is one tenth (of a meter), 10^{-2}, is one hundredth or 0.01, and so forth. On the right side of the circle, we descend into the micro world of ever-smaller structures.

The methods to probe these very small distance scales all rely on the basic fact that the resolution of a microscope is limited by the wavelength of the "light" one is using. As visible light has a wavelength of around

700 nanometers (1 nanometer = 10^{-9} meters), this sets a fundamental limit for ordinary optical microscopes. But in quantum theory there exist matter waves, the wavelength of a particle wave being inversely proportional to its mass times velocity. Thanks to this, we can reach very small wavelengths. An electron microscope is based on this principle and can go down to a scale of 0.1 nanometer or 10^{-10} meters: the size of an atom.

To reach even smaller scales, we must give the particles larger velocities, and that is exactly why we build particle accelerators. The higher the energy, the larger the velocities and the smaller the distances one can get a grip on. The smallest scales scientists have been able to probe that way is 10^{-20} meters, which is one hundredth of one billionth of one billionth of a meter! In big accelerators like at CERN in Geneva and at Fermilab near Chicago, physicists study the most elementary forms of matter and their interactions on that scale.

Our tour of scales has covered about fifty orders of magnitude and it is of course fascinating to think about the extreme sizes at both ends of the scale. But what is most remarkable is that at the human scale, of the order of one meter, we find ourselves pretty much in the middle. Over the centuries, we have managed to roughly balance our explorations of the larger and smaller scales. However, this may not be a coincidence ... Further on I will revisit this intriguing aspect of the scales of nature.

In exploring nature on all scales I have discussed the basic structures in which nature has organized itself. Now, there are four fundamental forces that manage to keep all those structures on all scales together. Only four forces? As far as we know, yes, only four. These four fundamental forces give rise to all interactions that take place in nature.

The notion of a *force* as such goes back a long way. Conceptually it was an important step. Originally, force was always presented in the context of motion. It is interesting to read Aristotle's account, because he claimed that in order for an object to move, a force was necessary. If one would stop applying the force, the object would stop moving. That may sound perfectly reasonable, because it confirms our daily experience, for instance when we move a book across a table. But perhaps you already know that the table also exerts a force on the book, a frictional force which counteracts the motion, and it is this force which causes the book to stop

moving. It took some brilliant thinking to figure out that the situation was basically very different from Aristotle's view. This was put forward by Galileo and later, in a very precise and general form, by Newton, who understood that in empty space, an object without any force acting upon it would keep moving indefinitely.

If a constant force is applied, like the gravitational force when we drop something, then the object will move with a constant *acceleration*. It was Galileo who made the crucial observation that the gravitational acceleration did not depend on either the material or the mass of the object. This is a special case of the key message of Newton's famous force law $F = ma$; if a force F is applied to a mass m, it will accelerate with an acceleration a. A special case of this law is the gravitational force near the earth's surface, where the gravitational acceleration is $a = g = 9.8\ m\ s^2$. Another special case of Newton's force law occurs when no force is applied at all: when

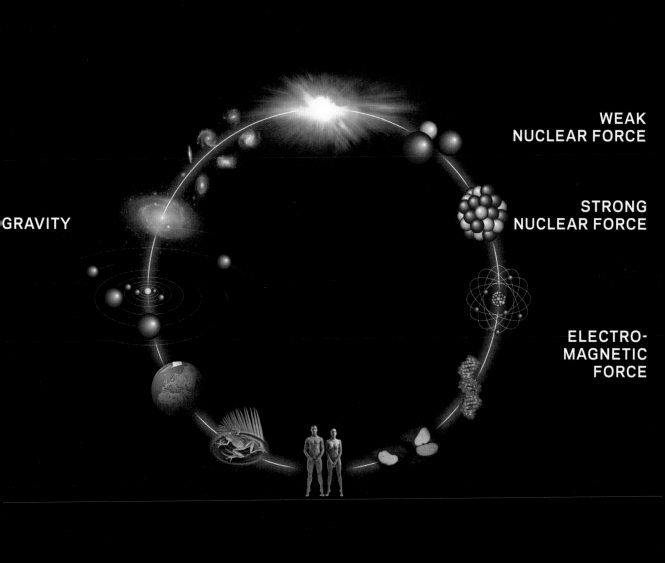

WEAK
NUCLEAR FORCE

STRONG
NUCLEAR FORCE

ELECTRO-
MAGNETIC
FORCE

GRAVITY

$F = 0$, then apparently the acceleration vanishes: $a = 0$. And indeed, no acceleration means constant velocity! (Which does not necessarily mean zero velocity, just that it does not change.)

The statement that we only know four forces may sound very strange to you, because forces are all over the place. You hear scientists talk of buoyant force, frictional force, muscular force, chemical force, molecular force, centrifugal force ... Not to mention market forces, peer pressure and The Force in *Star Wars*, which appear to be omnipresent as well. The explanation is simple: all real forces we know of in science are direct or indirect manifestations of the four fundamental forces.

In the previous section, you have seen how matter is put together in a very hierarchical way. So the question is simply: which forces hold the whole hierarchy of structures we discovered together? Two types of forces have been known since antiquity: gravity, and the electric and magnetic forces. The force of *gravity* is known to not only keep man tied to the earth's surface, but also to keep our solar system together, and the Milky Way, and the clusters of galaxies. In fact, that single gravitational force governs all large-scale structures and movements up to the scale of the universe as a whole – the entire left-hand side of our circle.

Now you might be tempted to conclude that it must be a tremendously strong force. Well, actually it isn't! Gravity is by far the weakest force we know. How do we know? Let's do an experiment. Put a nail on the table. Then take a one-dollar magnet, hold it above the nail, close enough so that the nail is attracted, and lift it from the table. End of experiment. What you have demonstrated is that the force exerted by all the matter of the earth pulling down on the nail is much weaker than the force the one-dollar magnet exerts on that nail. In a more refined thought experiment, it is possible to compare the gravitational and electric force that two electrons exert on each other. Electrons have a charge e and a mass m_e, they attract each other with a strength Gm_e^2/r^2 because of the gravitational force between them, and they push each other away with a strength fe^2/r^2 because they have the same charge, and like charges repel. The ratio between the strengths of the two forces at any distance is therefore given by Gm_e^2/fe^2, which can be calculated at 10^{-39}. This number shows how weak the gravitational force is with respect to the electromagnetic force, and as you can see it is an immensely small number. In other words, if one were to hold two electrons at a fixed distance and then let them go, they would fly apart immediately. Gravity would not manage to keep them together.

How can such a weak force keep a universe under control? There are various reasons. Firstly, the force of gravity works over unlimited distances, as far as we know, and secondly, there are enormous masses in the uni-verse. These compensate for the weakness of the force, because the gravitational force is proportional to those masses. And thirdly, we know of no other forces that work at such distances on neutral matter. So that is why I have placed the force of gravity on the left-hand side of the circle.

How different things are in the microcosm! We know that the electrons in atoms have a negative charge and the protons have an equal but opposite electric charge.[17] Unlike charges attract, and thus it is the *electromagnetic force* that keeps the atoms together. Because gravity is so much weaker, we may forget about it when talking about the forces in the atom.

Like gravity, the electromagnetic forces have an infinite range. Therefore they can be

[17] The choice of what is a positive or a negative charge is arbitrary and has a historical origin. There is no positive or negative connotation with respect to positive or negative charge.

observed at larger distances too. Electrical charge can be made visible by rubbing a piece of amber on a piece of wool and holding it above your head. Sure enough, your hair will stand on end. Magnetism was discovered because pieces of ore attracted each other. Even so, we notice relatively little effect of this force compared to the gravitational force. That is because all macroscopic matter around us is neutral: it has no net charge. If it did, we would notice immediately.[18]

We have now followed the right-hand side of the circle up into the nucleus of the atom. In that nucleus we find a bunch of protons sitting very close together. How is that possible? Like charges repel, and when they are very close together, they repel each other very strongly (because the force is inversely proportional to the distance). So these protons should fly apart if the electric force would be the only force we were dealing with. Yet they are tightly bound inside the nucleus, albeit together with a number of neutrons. The conclusion must therefore be that there exists another attractive force, which is stronger than the electromagnetic force. Indeed, the third force is called the *strong nuclear force*. It only works inside the nucleus – it acts at a very short range – and that is the reason why we didn't know about it until the first half of the twentieth century. Scientists had to discover the nucleus before they could find out about it.

So now we have: gravity – solar system; electromagnetism – atom; strong force – nucleus. What is the fourth force? It is called the *weak nuclear force* and its effect was discovered by Henri Becquerel and Pierre and Marie Curie towards the end of the nineteenth century, when they observed radioactive decay:

[18] From this we may conclude that the positive charge of the proton and the negative charge of the neutron have to be exactly opposite. The slightest imbalance between those two charges would upset the hierarchy of structures in nature.

the fact that the nuclei of certain instable elements can spontaneously change into other nuclei. In so called beta decay, a neutron is under the influence of the weak force converted into a proton, emitting an electron and a particle called an anti-neutrino. Clearly, this force works at the (sub)nuclear scale level. It is weak compared to the strong and electromagnetic forces, which work on the same scale, but is still much stronger than gravity. To return to all the other forces, the strong force primarily keeps the quarks together in the nucleus, but the quarks do not completely neutralize each other. The force that keeps the protons and neutrons together in the nucleus is nothing but the residual effect of the strong forces between the quarks in the different protons and neutrons. And something similar is true for atoms and molecules. The electrical charges are neutralized inside the atom, so the overall charge is zero, but because the plus and minus charges are spatially separated, there is a residual electric dipole force between atoms, which may bind them into molecules. A chemical bond can also arise because electrons orbit several nuclei in the molecule. And molecules in turn form greater structures or lattices, again based on the residual electrical forces. Once the structures become macroscopic, the gravitational force begins to dominate. But when you pull on a rope, for example, it is the inter-atomic electric forces that hold the rope together, allowing it to transmit the force you apply. And that is how forces on a larger scale arise out of the fundamental forces. So in the end all forces between matter can be accounted for by a combination of the four fundamental forces we have discussed. The modern description of a force is based on the notion of a particle that is exchanged and carries the force. For example, the electromagnetic force is carried by the photon, which if exchanged between two electrons will lead to the repulsion between the two.

Our next overlay of the circle shows a kaleidoscope of the sciences. At each scale on which nature has organized itself into a typical structural entity, you see a matching field of science. This is a bit like in the Middle Ages, when towns typically developed at the crossings of rivers and roads. Each scientific field developed rather autonomously, with its own instruments, units, jargon and unsolved problems. And all these fields are still developing: they have ways to go to completely fulfill their mission.

I can only give a rather coarse-grained view here, so if you miss your favorite science – be it laser science, biochemistry, meteorology, linguistics, neuropsychology or the search for extra-terrestrial intelligence – please imagine it falling under one of the other headings. Indeed a much finer mesh of scientific disciplines could be placed on the circle, and with all the modern interdisciplinary fields like astro-biology, psycho-chemistry and so on, it would become very crowded indeed.

Let us look at the sciences on the circle. At the very smallest scales on the top right we have the field of *elementary particle* or *high-energy physics*, in which the smallest, most elementary forms of matter and the laws that govern their interactions are investigated. Moving down the circle and up in scale, you see *nuclear physics* – the physics of nuclear energy, the study of the properties of nuclear matter. One of those properties is the nuclear magnetic spin, which is probed in magnetic resonance imaging devices. One more step takes us to the science of the *atoms*, which corresponds with the more than hundred chemical elements appearing in nature and which are organized in the periodic table of Mendeleyev.

Then we arrive at systems that are composites of some or of very many atoms. We enter the field of *chemistry*, which deals with the incredible variety of molecules that may be formed. Of particular interest is *organic chemistry*, centered around evermore com-

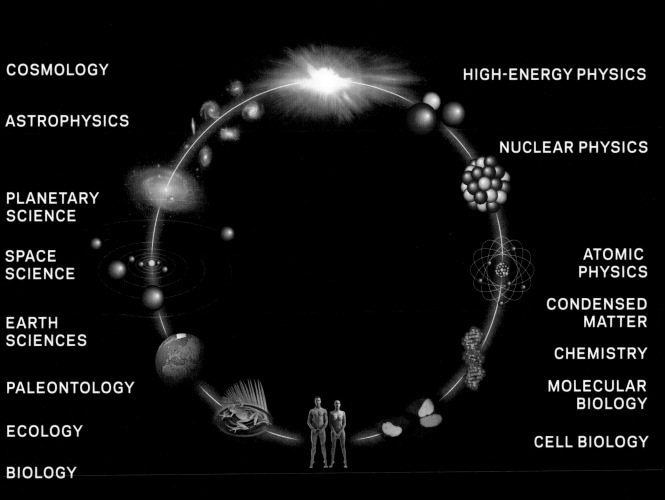

COSMOLOGY

HIGH-ENERGY PHYSICS

ASTROPHYSICS

NUCLEAR PHYSICS

PLANETARY
SCIENCE

SPACE
SCIENCE

ATOMIC
PHYSICS

CONDENSED
MATTER

EARTH
SCIENCES

CHEMISTRY

PALEONTOLOGY

MOLECULAR
BIOLOGY

ECOLOGY

CELL BIOLOGY

BIOLOGY

MEDICAL SCIENCES

plicated and larger molecules composed of large numbers of carbon and hydrogen atoms, such as polymers and buckyballs. The molecular world extends all the way up to *biochemistry*, the chemistry of the molecules of life, like proteins, DNA and RNA.

Still, molecules are not the only way in which many atoms can organize themselves under the influence of the inter-atomic forces. There is also the tremendous structural diversity of *condensed states of matter*, such as fluids and solids. Think of the highly symmetric crystals in all those beautiful minerals and gems, but also of amorphous materials such as glass. Think of semi-, super-, and ordinary conductors, and of insulators. Think of liquid crystals like the ones used in digital displays and of nanomaterials with their promise for new technologies.

If again we move one step up in scale, we enter *molecular biology*, the study of the behavior of the molecules of life. We know that DNA with its roughly 3 billion base pairs forms the molecular expression of our genetic code. As we are now learning, the DNA contains about 22,000 genes that code for the proteins that allow our bodies to function. These proteins function independently, but they also work together in networks of varying complexity. How they manage to run a biological organism is still largely unknown. Between the genetic code and the cell or the organism, there is the vast domain of genetic expression and regulation. *System biology* studies the whole of chemical and biological activities and functions in the living cell, which in turn is the universal building block of all more complex organisms. Biologists are discovering a whole tower of epigenetic organizational levels that regulate the genetic expression in the more complex forms of life. Understanding these networks and hierarchies is one of the great challenges in modern science.

The advance of molecular thinking has opened the door to a revolutionary new approach of living organisms. And that is how we arrive at the human figures at the bottom of the Ouroboros, symbolizing the most complex organism we know.

Now let us examine the other half of the circle. At the very largest scales near the top we have *cosmology* and *astronomy*, nowadays pursued through a host of observatories in outer space such as the Hubble space telescope. It allows us to look deeper into the universe than we ever did before and provides us with spectacular images of gravitational lensing, the remains of supernovas, colliding galaxies, and so on. Another example is the WMAP satellite, through which scientists observe irregularities in the microwave background radiation that fills the universe. These measurements provide us with important clues about the energy composition and distribution of the universe and thus about the various stages of cosmic evolution. Exciting too is the search for *exoplanets* - planetary systems other than our own.

This search started in 1994 and has now produced a catalogue of over 300 such systems in our galactic neighborhood. The old question whether extra-terrestrial life exists becomes more pressing than ever!

We move down in scale and closer to home from research on the scale of galaxies, via the study of neutron stars, white dwarfs, black holes and other exotic inhabitants of space, to our familiar solar system and the planets orbiting the sun – and one of them in particular: our own earth.

Research in the *earth sciences* takes place on many levels, starting with the overall scale. Firstly there is the field of geology, concerned with both the surface and the interior of the earth. Most of the earth's geological structures, such as the positions and origin of mountain ranges, and geological dynamics like earthquakes, tsunamis and the occurrence of volcanic activity can be understood in terms of the theory of *plate tectonics*. The tectonic plates are driven by slow currents in the layer of molten rock underneath them. Along their boundaries these plates may diverge, collide, or shift alongside each other, and such motions can cause disruptions leading to large-scale catastrophes. The earth sciences also include the study of the seas and climate, where researchers strive for a good understanding of such vital issues as global warming and the rising of CO_2 emissions. These studies may be of tremendous importance for our own immediate future, and even more so for the generations after us.

Moving one more step down, we encounter – but now from the macroscopic side – the phenomenon of life on earth. Through *ecology* we come to *biology* as the study of all living species and their evolution, as envisaged by Darwin.

Having moved down from the largest scales we encounter life, like we did by moving up from the smallest scales. In the phenomenon of human life, the icon of complexity, large meets small. Scientifically speaking the human being is still largely unknown, a *persona incognita*. Health and illness of this unique organism are studied in the *medical sciences*. A growing variety of fields clustered under the denominator of *life sciences* – psychobiology, cognitive sciences, genetic engineering, *et cetera* – try to create order by careful observation, rigorous data analysis and sophisticated mathematical analysis. These efforts may ultimately provide us with the understanding of who we are and how we function – all the way, from the organs to the immune system and our brain. In this ultimate scientific challenge, man confronts himself as "just" another natural phenomenon. Beyond the study of the self as in psychology, lie the questions of collective behaviour as studied in sociology and anthropology, but also in social geograph and economy. All these questions beg to be understood on a firm scientific basis both conceptually and quantitatively.

I have provided a brief general overview of the circle of sciences which surrounds the mysteries of nature. I would like to mention that of course mathematics and informatics have been indispensable for the achievements made in all these fields of science.

In spite of their great successes, all branches of science struggle with their own "favorite" unsolved problems – whether it is the big bang, the switching of the earth's magnetic field, the meaning of junk DNA, the laws of quantum gravity, the origin of life or the workings of our brain. Science is always immersed in a sea of questions waiting to be answered, if not today by our brains, then tomorrow by the brains of our children. Whether at some point those brains will still be human or somehow artificial, that doesn't really matter. Nature is impressive in its diversity and beauty, but perhaps most impressive of all is its incessant quest for understanding itself through the evolution of consciousness and the human mind.

Though this book is primarily about science and its meaning for humanity, in Chapter II I have already emphasized the fact that science and technology are intimately linked. They are indispensable for each other and push each other forward in the double helix of knowledge and technology. It is a perpetual waltz of observation with imagination, of mind with matter, of the quest for truth with the quest for new inventions. Therefore I present an overlay of the Ouroboros with a wild scattering of technologies that have originated all along the circle.

Technologies have changed our daily lives in unprecedented ways. New technologies are the dominant driving forces behind economic and social progress. At the same time, we should be aware that in the hands of society technology is a double-edged sword. because it has a constructive as well as a destructive potential.

Its dark edge is most obvious in the needless wars fought using the most atrocious weaponry, but also in the excessive forms of consumerism created by a concentration of too much power on the side of production and media conglomerates. We have as a society to cope with military industrial and medical industrial complexes but also with food and fashion addictions that do not stand out as pinnacles of civilization. The sword's light edge shows itself in the overall emancipation of mankind: the liberation from ignorance, exploitation and irrational fear. We are faced with the tremendous challenge of turning the enormous potential of technology into a true improvement of the human condition.

To achieve that it does not suffice to educate concerned scientists and engineers: what is needed is a broad awareness of this key message in our political and economic institutions and our everyday media. After all, they, more than anybody else, are responsible for the smooth implementation of new technologies to the benefit of society. While the title of my book *In Praise of Science* can

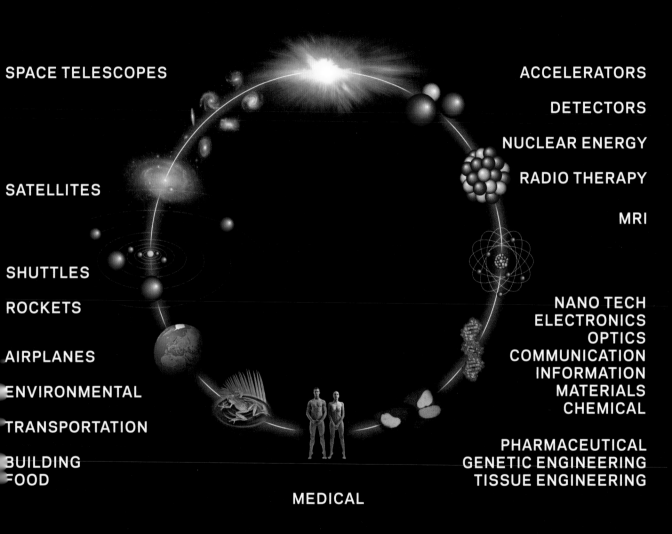

SPACE TELESCOPES

SATELLITES

SHUTTLES

ROCKETS

AIRPLANES

ENVIRONMENTAL

TRANSPORTATION

BUILDING
FOOD

MEDICAL

ACCELERATORS

DETECTORS

NUCLEAR ENERGY

RADIO THERAPY

MRI

NANO TECH
ELECTRONICS
OPTICS
COMMUNICATION
INFORMATION
MATERIALS
CHEMICAL

PHARMACEUTICAL
GENETIC ENGINEERING
TISSUE ENGINEERING

be read like *The praise of knowledge* which is easy to defend, that would have been much harder if the title had been *In Praise of Technology.*

I have placed a rather arbitrary variety of technologies around the circle. Because you already know most of them, it doesn't serve any purpose for me to discuss them in great detail. Technologies also form some an evolutionary tree, with very elementary inventions at its roots, like the making of fire the invention of stone tools or the invention of simple devices like the lever and the wheel. These marked the basic awareness of the importance of shaping and "processing" materials to extend their use, and the importance of energy and of saving energy by constructing simple machines that extended the physical capabilities of the human being. Even in antiquity many civilizations developed technological capabilities that even impress us now. The construction of very large temples, pyramids and palaces, but also of ships, irrigation works, roads and large cities all involved great mastery of a great variety of design and technical skills.

From simple machines exploiting just human force, a major breakthroughs came when we learned to use other energy sources to power them, wind, water and steam pressure. The industrial revolution showed an impressive growth of production through a large scale mechanization.

A new impetus was the ability to generate and distribute electricity and to convert it to mechanical power with the electromotor. Industrial processes became more and more complex and automated. Physical labor became more and more the task of machines. With the discovery of electromagnetic radiation, and inventing ways to generate such radiation for example radio waves, global communication became feasible.

Then we made the transition from electric devices tot electronics, first with the use of radio tubes but later by developing semiconductor technologies. That way a radically new technological dimension opened up in the computer era. Computers immensely extended the human capacity to store, process, and exchange information. So, after the periods of mechanization, and automation we entered the modern information era, with laptop, Ipod, and Internet as the icons of modern times.

Perhaps it is worth just spending some time staring at this circle. What did it look like for your grandparents or you great grandparents? And what will it look like for your grandchildren? Ask yourself if you can add five more important technological breakthroughs that are not listed. Observe how technologies often involve highly integrated efforts of different parts of science. Airplane design, for instance, makes use of knowledge of aerodynamics, fuel combustion, communication technologies, but also of new carbon fiber materials.

The collaboration of many hands and brains is turning our world into a global village, with one global climate system, one global economy, one Internet and hopefully at some point one global sense of justice.

I believe that the spiral of science and technology has changed the world more profoundly and irreversibly then any economical, religious or political doctrine. And this process will continue involving future generations, at least if we manage to make it more refined, more respectful of nature, and less wasteful and destructive. Technology has taught us many things, but one of the most important lessons is how fragile our environment is, and we ourselves. Today, the central question to address is how we can force our technological society through a transition towards a full commitment to the creation of a more sustainable world, where our grand grandchildren will enjoy at least the same basic resources, goods and opportunities like we did. This would be a tremendous turning point, and once this turn would be taken it would clearly generate ample high and low tech economic opportunities. We have to push ourselves over that bumb, that's all. A bright future is at least still possible.

In the previous overlays we have zoomed in on specific sciences and technologies. Let me now do the opposite and zoom out as far as possible, to look at the Ouroboros as a whole again. In the illustration on the facing page you see the three ultimate frontiers of our knowledge: the very large, the very small and complexity.

As I mentioned before, all fields of science have their great unknowns. But if you look at science as a whole from a distance, you see that our drive for knowledge has always been pushing in three directions towards entirely unknown territory. When we get results from a new accelerator, that is comparable to Van Leeuwenhoek looking through his first microscope and observing things that nobody had ever seen. A more objective step towards new knowledge is hard to imagine.

Pushing for the ever smaller, we now wonder whether quarks and electrons are the smallest entities in nature. Should we per-haps anticipate much smaller objects, such as superstrings? It is possible that on those tiny scales spacetime has many more dimensions than the four we can observe now – in fact, string theory predicts this. You may not worry about such outlandish matters, but many physicists do. The true number of dimensions of spacetime being ten or eleven is an exciting idea. It may seem inconceivable now but perhaps in fifty years time this will be common lore, to be answered with a mumbled "of course."

The fundamental question is whether there will always be smaller substructures, like Chinese boxes or Russian Matryoshka dolls. Is there an infinite hierarchy or will it stop, is there a smallest scale where it all ends? That is definitely a logical and physical possibility; it would imply that once we have arrived "down there," nothing new will be found. The only activity left would be scientific tourism for fun, for inspiration or out of nostalgia. But fortunately it is too early to make such

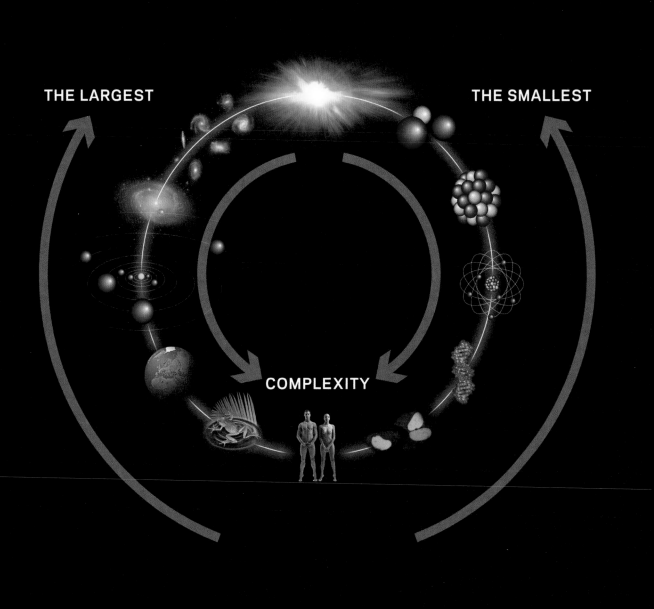

THE LARGEST

THE SMALLEST

COMPLEXITY

bold statements. New data may shatter our expectations and lead us in a completely different direction, as new data have done so often before.

A similar story can be told about the large scales, as we approach the boundary of our universe. Is there indeed such a boundary, or might we enter regions of the cosmos where the universe is very different from ours? Some scientists believe the cosmic landscape predicts such a diversity, a multiverse filled with pocket universes, each with its own variety of physics – and only a few of them inhabitable.

Interestingly, these questions of the smallest and largest are related. I will come back to this topic shortly, but let us first have a look at the two arrows pointing downward towards complexity. There are two arrows because one comes down from the macroscopic side – the Darwinian side of evolution – while the other comes down from the microscopic side – the side of molecules and cells. The word complexity stands for an enormous array of poorly understood phenomena, like weather systems, genetic trait development, food webs in ecology, financial markets, and the brain. Some say such processes are poorly understood exactly because they are too complicated, which seems to me too simple an explanation. Others say that it takes another way of looking, other tools of observation, or different mathematics to crack the central problems of complexity. This view places complexity back into the realm of the familiar sciences, where the instruments, the mathematics and the fundamental concepts have co-evolved with the objects of study. These are developments that are now taking place at institutes like the Santa Fe Institute.

There are various levels on which you can try to model social dynamics. So far, the study of very complex systems has led to theories of surprising simplicity, often even of some superficiality. This would be great if they would be precise and would lead to reliable predictions. This is mostly not yet the case and that points to the fact that most of these sciences are very young. To a large extent their progress is directly linked to the availability of large computer facilities, allowing researchers to analyze huge amounts of data and to run large-scale simulations of very complicated model systems. Internet machines like Google are clearly a very rich source of reliable and detailed data, which could be most relevant for the development of a more quantitative approach to the social sciences. These fields look extremely promising. As their diagnostic tools improve, so will their ability to probe vast amounts of reliable data on all scales, and a precise understanding of complexity will come within reach.

There is one more aspect to be noted about the three frontiers. As you can see, the two upward arrows roughly mark the history of human discovery, with the frontiers of perception shifting ever more outward and inward. On the other hand, the downward arrows, marking a growing complexity, represent the history of the universe itself. I will clarify this further on.

The most important task for scientists is to search for the most fundamental laws from which a picture of the world can be deduced. There is no logical path that leads to these elementary laws, only an intuitive one, based on creativity and experience.

With such a methodological uncertainty one would think that an arbitrary number of equally valid systems would be possible. However, history shows that of all conceivable constructions always a single one did stand out as absolutely superior to all others.

Albert Einstein, 1918

There are different kinds of great achievements in science. We may make groundbreaking observations, or discover crucial correlations between different observations. We may develop models which lead to a thorough theoretical understanding, or even to a set of principles allowing for a causal explanation of a huge class of observations. When that happens, a big research effort, extended in time, can condense into a compact and powerful statement about how nature works. Such a discovery need not be instrumental for technological progress; I am not talking about something like the invention of the wheel, but rather about a revelation like the general laws of motion. Not about the discovery of neutron stars, but about the laws of relativity of spacetime.

I refer to such momentous steps as "turning points." You might say: why not call them revolutions? They certainly were, in that they overthrew our common understanding of things. But I prefer to call them turning points, because the word revolution has the negative connotation of destruction. Thomas Kuhn used the term "paradigm shifts," indicating that these moments require changing our mental framework. You see, I am really referring to turning points in our thinking. To begin with, we are all working and thinking in one direction, perhaps without even being

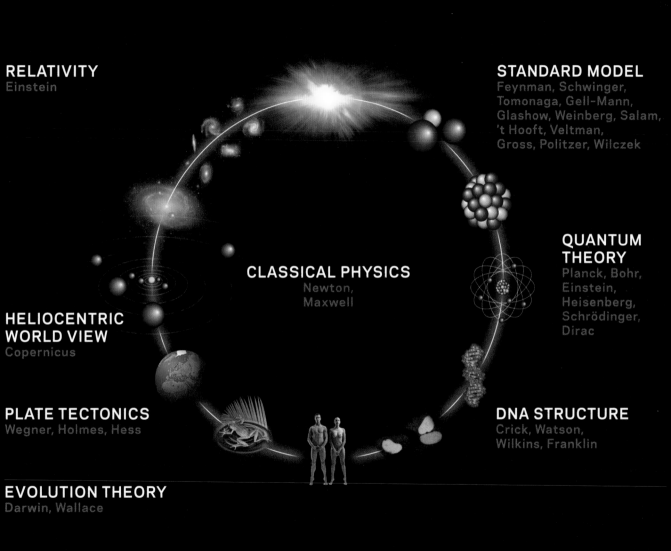

RELATIVITY
Einstein

STANDARD MODEL
Feynman, Schwinger,
Tomonaga, Gell-Mann,
Glashow, Weinberg, Salam,
't Hooft, Veltman,
Gross, Politzer, Wilczek

CLASSICAL PHYSICS
Newton,
Maxwell

QUANTUM
THEORY
Planck, Bohr,
Einstein,
Heisenberg,
Schrödinger,
Dirac

HELIOCENTRIC
WORLD VIEW
Copernicus

DNA STRUCTURE
Crick, Watson,
Wilkins, Franklin

PLATE TECTONICS
Wegner, Holmes, Hess

EVOLUTION THEORY
Darwin, Wallace

aware of it, and then more and more anomalies are found and we slowly begin to realize that the theory is never going to work. Then somebody comes by and opens this door that nobody had even noticed; a door to something completely new. As Einstein once remarked, real problems can rarely be solved at the same level of thinking at which they were created. That describes the situation well.

Turning points in science are precise statements about how nature functions. That does not mean they express some absolute truth, or anything like that. They are just exceptionally robust statements that may go unchallenged for hundreds of years, and when they are challenged, it is mostly in order to be absorbed in a more unified overarching theory or framework, which shows their limited range of applicability. For instance, Newton's laws were not really "wrong," they simply turned out to have a limited range validity. Even after the discovery of relativity, in which Newton's laws were replaced by a new, conceptually very different set of equations, the old theory remained a fantastic approximation to reality, but in a restricted domain. Even today, 99.99% of our daily-life mechanical technology relies on Newtonian physics. Usually, bridges do not collapse and planes do not fall out of the air, and if they do, it is never Sir Isaac Newton who is to blame.

The opposite may also happen: sometimes only long after its conception the full meaning of a theory is appreciated. I am thinking for instance of the theory of nonlinear dynamics and deterministic chaos, which can already be found within the framework laid out by Newton three hundred years earlier. Another example is the advent of quantum information theory, some 75 years after all the necessary concepts were written down in painstaking detail by Schrödinger, Heisenberg and Dirac.

According to me, such turning points should be recognized as key events in the evolution of human thought. They are not just some freaky science thing that only nerds can get excited about. And that is why I will discuss the turning points shown on the Ouroboros in some detail.

Let me start with what is now considered to be the beating heart of *classical physics*: the laws of mechanics and the fundamental equations of electromagnetism. *Isaac Newton* was the first to give a very precise mathematical description of velocity and acceleration. It allowed him to formulate the three laws of motion, like the force law $F=ma$, that are still taught in all schools today. He also added a fourth law, which is his gravitational law. This simple law states that the force with which two masses m_1 and m_2 attract each other is proportional to both masses and inversely proportional to their distance squared. This force law used to be called universal, as it held in exactly the same way for a pen dropping onto the floor as for the moon orbiting the earth or the earth orbiting the sun. Newton brought the terrestrial and heavenly mechanics together in one set of universal laws, formulated in an exact language of mathematics that he himself developed – an absolutely magnificent accomplishment by one of the most extraordinary geniuses in the history of science. The laws provided the fundamental foundation for *Copernicus'* description of the solar system, as well as for *Keppler's* laws of planetary motion.

But at the time of the Enlightenment, the gravitational force was not the only force identified by scientists. The electric and magnetic forces were also begging to be understood. After a long build-up of many remarkable observations and partial explanations, around 1865 the theory was finally completed in a brilliant way by *James Clerk Maxwell*. His description brought together all electromagnetic phenomena in just four equations. And as a bonus these equations explained the phenomenon of light as an electromagnetic wave propagating through spacetime, its wavelength (or frequency) determining the light's color. It was already known that all types of light moved with the same velocity, which had been measured to be close to 300,000 km/s. Maxwell's equations are a good example of a tremendous unification in our understanding of different physical phenomena. Where once electricity, magnetism and light were considered to be entirely disconnected phenomena, they were now brought together under one umbrella.

I have placed these two great turning points in the center of the picture, as they represent the first convincing evidence that our universe was obeying universal laws, which could be discovered through careful observation and, accurately described in the language of mathematics. After these achievements, at the end of the nineteenth century, many physicists were convinced that their field of study was nearing its completion: it was just a matter of getting the last de-

tails right. Yet not long afterwards many of those very same physicists had to conclude that their premature judgment had been all wrong.

During the first quarter of the twentieth century, there were some enormous conceptual transformations, caused by a number of giant blows to the admirable edifice of classical physics. Two of them dealt with the most basic concepts of space and time: the *Special Theory of Relativity* of 1905 and the *General Theory of Relativity* of 1915. Both were conceived by *Albert Einstein*, who was the first to take the very bold steps necessary for an entirely new view of our universe.

His first step was to introduce the notion of a four-dimensional *spacetime*, having three space and one time dimension. Relativity refers to the fact that our state of relative motion determines how the space and time axes are oriented in spacetime. It meant that time lost the absolute status it had enjoyed since Newton: it now became relative, just like space. The status of absoluteness was instead granted to the velocity of light. Firstly, this velocity would be the same for all observers. And secondly, nothing could go faster than the speed of light. $c = 300,000$ km/s is the absolute speed limit; no speeding possible! In Einstein's world view, observers moving at large velocities with respect to each other may find that their clocks run at different rates. This evokes incredible effects like the twin paradox, which states that if one of a pair of twins goes on a long, fast journey through space, upon returning home she will find her sister to be much older, if not deceased! That sounds weird. But the effect is true, and moreover it has been experimentally confirmed – not by sending twins to Andromeda and back, but rather by studying the velocity dependence of the lifetime of unstable particles in a lab.

The radical reinterpretation of space and time had far-reaching effects. It led to the insight of the equivalence of mass and energy, epitomized by the powerful equation $E = mc^2$, known to many but understood by few. And if mass is just a form of energy, that poses the challenge to liberate that energy. This story, as you know, has a tragic plot line in the development of atomic weapons and a positive one in the peaceful applications of nuclear energy.

The uses of atomic energy illustrate an essential aspect of the turning points: in and of themselves they are victorious moments in the history of understanding, but in their applications the new insights invariably show two faces. The dark one might involve a military invention, while the bright one improves the human condition in some essential way, for example by providing a new diagnostic tool.

After almost ten years of solitary explorations, trying to extend his Special Theory of Relativity to accelerated observers, Einstein made the startling discovery that gravity could be interpreted as the "curvature of spacetime." This new theory of gravitation had very significant consequences that nobody had anticipated. Einstein's equations linked the local curvature of spacetime at any point to the energy and momentum locally present in that region. They led to seven predictions that have all been verified by experiments, although some of the effects are very small and hard to measure directly. The most striking consequence is that the universe as a whole had to be dynamic. The equation basically predicted that our universe must be expanding, which was confirmed by Edwin Hubble in 1928.

Now have a look at the other side of the Ouroboros. At the micro scale, the cherished laws of classical physics were to be dethroned in an even more radical sense. The classical laws of mechanics and electromagnetism appeared to fail completely. In 1911 Ernest Rutherford had made the discovery that atoms were not just neutral homogeneous chunks of matter. On the contrary, he found that all of the atom's mass was basically centered in its center in a positively charged nucleus, while very light, negatively charged electrons circled around the nucleus at a distance. Initially, these findings seemed to fall entirely within the reach of classical physics. But Maxwell's equations predicted that the electrons would lose energy by emitting electromagnetic radiation, which would cause the electrons to fall towards the nucleus. This would happen within a microsecond, and therefore atoms could not exist according to the laws of classical physics.

Such a total crisis calls for an extremely creative move. The creative move came from the minds of the first generation of *quantum physicists*, *Max Planck*, *Niels Bohr* and again *Albert Einstein*, responding to a number of inconsistencies. *Werner Heisenberg*, *Erwin Schrödinger*, *Paul Dirac* and others transformed their somewhat arbitrary rules into a rigorous theoretical framework.

The turning point of quantum theory led to

astounding new insights which surprised everybody, including the inventors. It brought us the complementary view of particles and waves. Particles could behave like waves, for example by showing interference patterns, and waves could behave like particles. Light waves, for instance, could be interpreted as a stream of light particles or photons, that carried a well-defined energy and momentum just like other particles. But waves are not point-like: they never exist solely in one spot. This paradox led to the formulation of the *uncertainty relations*, one of the more abstract but very fundamental principles of quantum theory. Physicists were confronted with a theory whose only consistent interpretation at the most fundamental level was a probabilistic one.

Other strange laws were discovered, such as the Pauli exclusion principle, which stipulated (i) that particles could only belong to one of two classes, bosons and fermions, and (ii) that two identical fermions, such as the electrons in our atom, could not occupy the same state. This principle has immense consequences: it makes it possible to understand many properties of matter, not only at the atomic scale but also on a macroscopic scale. It is the basic law behind the periodic table: because electrons cannot all occupy the lowest energy state, they have to successively fill the higher levels. This causes the atoms' very different chemical properties. The Pauli principle also explains why conductors exist: in these materials, the atoms' electrons are forced to occupy high-energy states in which they are loosely bound and thus free to move around.

Quantum theory truly opened the doors to the microcosm, and its principles are still valid at the smallest scales being investigated today. I have mentioned the nuclear forces that work at very small distances only. The turning point related to them is called the "*standard model* of elementary particles and forces." It is in fact a quantum theory describing all known fundamental particles, like quarks, electrons and neutrinos, and the fundamental forces between them – except the gravitational force. This is what is often called a "unified field theory" because it unifies the detailed description of wildly different physical phenomena at the (sub)nuclear scale.[19]

You can see that the principles of quantum theory govern almost the whole right side of the circle: from particle physics all the way to chemistry. The technological consequences of quantum theory have been enormous, too, from semiconductor physics (i.e. computers) to lasers, nanophysics and now quantum information technology.

Let us now turn to the bottom-left part of the Ouroboros. The first turning point to mention here is the theory of plate tectonics, which is the central dogma of modern geolology. It grew out of *Alfred Wegener's* original conjectures about *continental drift* around 1915. The theory of plate tectonics is connected with the names of Arthur Holmes and Harry Hess. It was firmly established around 1960. Deep-sea measurements of the magnetic field at the earth's surface could be correlated with the periodic inversions of the earth's magnetic field, providing very strong and detailed evidence. There are eight large plates in the outer layer of the planet, as well as many smaller ones, moving with respect to each other at typical velocities of a few cm per year (about the speed of a growing nail). The driving force is provided by the convection currents in the deeper layers of the earth. The theory explains many geological events, like volcanic activity, mountain range formation, earthquakes and tsunamis. Most of these originate along the boundaries where the plates move with respect to each other.

[19] This theory was developed over a period of about 40 years and many names can be associated with it, we only have indicated a few of the most prominent ones.

The next turning point, *evolution theory*, is of great depth and has had tremendous impact. In *On the Origin of Species* (1859), *Charles Darwin* gave his grand view on all life on earth. He brought about a huge change in people's way of looking at the bewildering beauty and diversity of living beings, by adding a time axis to the frame of reference. By placing the living species along a time axis, he gained the insight that these organisms could have evolved out of other more primitive ones. It was a move from a view of nature as a very complex stationary state to one in which nature was viewed as a dynamical process.

Finally, the several classes of facts which have been considered […], seem to me to proclaim so plainly, that the innumerable species, genera, and families, with which this world is peopled, have all descended, each within its own class or group, from common parents, and have all been modified in the course of descent, that I should without hesitation adopt this view, even if it were unsupported by other facts or arguments.

Charles Darwin, 1859

And that was not all. Darwin also described the mechanism by which the evolutionary process took place: variation of properties and natural selection of the individuals most fit to adapt to the given situation. Again the consequences of the insight were far-reaching – as we know, because 150 years after the theory was formulated, there is still a fierce debate about its validity and meaning, if not among scientists then certainly among the general public, as I have pointed out in the first chapter. It is not easy to cope with the fact that your ancestry comprises primates, fish, and ultimately protozoa of some sort. Evolutionary thinking has deeply influenced the modern social sciences, and even fields like economics. It has been said that evolution theory is a qualitative mechanism, rather than a law like the ones I discussed before. The theory does teach us that in addressing problems it can help to think of processes instead of situations. And it also teaches us that very simple rules or algorithms can give rise to very complicated patterns. Evolution theory does not stand out because of its predictive power, but it allows us to recognize the importance of principles like *fitness*, *contingency* and *emergence*.

The notion of emergence refers to the property that the collective may exhibit properties that the parts themselves do not have. So the whole can be more than the sum of its parts. One plus one is three. It is crucial to realize that these properties are not added by hand, they emerge in the collective. A simple example is a water wave, which is a property that a large collective of water molecules can exhibit. But the notion of a wave doesn't make sense for a single molecule.

The notion of contingency refers to the acci-

dental nature of the final state, *i.e.* we might not have been here if not about 60 million years ago a huge asteroid would have hit the earth near Yucatan, thereby disturbing the biosphere in such a dramatic manner that the dinosaurs went extinct.

This is an event that clearly had great impact, yet was accidental. It is the well-known "if": what would have happened if Napoleon had died from pneumonia at an early age? What would the world look like if we run the tape of evolution again starting at the same initial condition? What it really means, is that the outcome of evolutionary processes is in principle dependent on the details of the history along the particular path. And that again makes it hard to model, because the path is through a space of possible occurrences about wich we do not know too much.

Fitness refers to the dynamics of natural selection: it is the notion of being well adapted for a given situation or challenge; in evolutionary biology it is quantified in the reproductive success rate in a genetic sense.[20]

These notions are crucial in the modern sciences of complexity, where one thinks in terms of clustering, adaptivity, network dynamics, and agent-based modeling.

Networks have vertices and links, the vertices have certain properties represented by a set of numbers. The links represent the interactions that also may be quantified as being strong or weak. You must now imagine that this network gets permanently updated according to some algorhythm, which produces the new values in terms of the old ones. Now it may be that statistically some equilibrium or stationary state arises in which the distribution of values over the collective doesn't change anymore. How dependent is the resulting distribution on the inputs or on the rules, on the strength of the interactions

and so on? These modeling techniques allow you to study the aforementioned biological properties.

Some networks are not passive but active, consisting of many agents obeying simple local rules, and determining certain strategies using game theory. In these fields the emergent properties of the collective are then studied using large scale computer simulations. These simulation techniques make it possible to model phenomena which are hard to narrow down to a few variables and parameters, systems in which many different spatial and temporary scales are simultaneously relevant to find out about the robust or resilient properties of the network as a whole. Such modeling techniques for complex systems do not only find applications in the arena of biology, but also in modeling social networks: belief systems, decision networks, models for disagreements, and in the economical arena, for example in the field of econophysics.

[20] It is the ratio of the number of individuals carrying your genetic code before and after natural selection has taken place. Fitness therefore depends on time.

Biology has seen a remarkable turning point on the microscopic side as well. Microbiology moved from biochemistry to molecular biology, now one of the largest fields of scientific research. This field has seen many drastic changes in a short period of time, but I have chosen the discovery of the *structure of DNA* by *Francis Crick* and *James Watson* (and, often not mentioned, *Rosalind Franklin* and *Maurice Wilkins*) as the turning point. Thanks to this discovery, very precise thinking about life from a molecular point of view became feasible. It really turned biology into a hard science, so to say, in spite of the complexity of the subject. And once we look on the level of the rather independent building blocks of life, there is more than DNA alone. A perplexing complexity of functions and actions is performed by a very complicated interplay of proteins, RNA molecules and networks of them. For me, perhaps this is where the most exciting science is going on right now, as researchers attempt to fully understand the complete chemical processes taking place in the cell.

It is exciting to notice the wonderful complementarity between the macroscopic neo-Darwinistic approach to life and the microscopic molecular approach based on the genetics encoded in the DNA molecule. Paleontologists gain new information from researchers studying the similarities of genetic material of the various species, while at the same time they provide molecular biologists with new clues on which links to search for.

Have there been no more turning points in the last 50 years? My view is that we are banging on the doors of complexity and that we have learned a great deal in this area. Researchers have gathered tremendous amounts of information, and found many suggestive correlations in a wide range of fields. They have come up with all kinds of partial explanations on different levels. Yet my impression is that so far we have not yet seen a turning point, a change of thought as essential and far-reaching as the idea of evolution or relativity or the structure of DNA. But then, those were not anticipated either, so big surprises may be in store for us.

The turning points tell a fantastic story of the human mind, its inquisitive nature, its endurance, and its promise. The story captures our history in a way that is different from the way world history is usually told: it does not include Napoleon, Queen Victoria, Roosevelt, or Mao Tse Tung. It is the story of liberating our minds to the extent that we can face the amazing challenge of exploring our own origins and shaping our own future.

There was not then what is, nor what is not. There was no sky, and no heaven beyond the sky. What power was there? Where? Who was that power? Was there an abyss of fathomless waters?

There was neither death nor immortality then. No signs were there of night and day. The ONE was breathing by its own power, in deep peace. Only the ONE was: there was nothing beyond.

Darkness was hidden in darkness. The all was fluid and formless. Therein, in the void, by the fire of fervor arose the ONE.

And in the ONE arose love. Love the first seed of soul. The truth of this the sages found in their hearts: seeking in their hearts with wisdom, the sages found that bond of union between being and not being.

Who knows in truth? Who can tell us whence and how arose this universe? The gods are later than its beginning: who knows therefore whence comes this creation? Only that god who sees in highest heaven: he only knows whence comes this universe, and whether it was made or uncreated. He only knows, or perhaps he knows not.

Rig Veda x.129
translated by Juan Mascaró

I have discussed the gigantic hierarchy of stable structures that make up what we call nature. We have also looked at the conceptual framework in which these structures can be understood. And now I will address the simple question: where do these structures come from? Were they all present already at the time of the big bang? Or did they somehow come into being, in some particular order? If so, can we then understand how that came about? This is the basic question about our cosmic origins. The answer lies to a large extent in Einstein's theory, which describes the big bang and the subsequent expansion of the universe. I will describe this process while referring to our Ouroboros, in which I have used arrows to indicate the sequence of structural

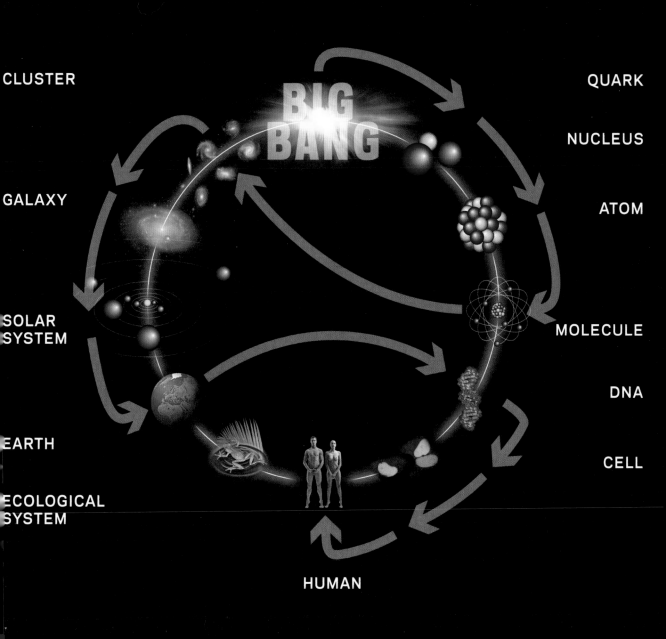

CLUSTER

QUARK

GALAXY

NUCLEUS

ATOM

SOLAR
SYSTEM

MOLECULE

DNA

EARTH

CELL

ECOLOGICAL
SYSTEM

BIG
BANG

HUMAN

development starting from the big bang.

At the time of the big bang, the universe was extremely dense and extremely hot – in fact, almost infinitely dense and hot. It was a sort of boiling soup of the most elementary forms of matter, such as *quarks*, *electrons*, *photons*, *neutrinos* and the like, as well as their anti-matter counterparts. Because the universe was expanding, its temperature dropped and therefore the average energy with which the particles collided went down. This meant that it became possible for the first bound systems to form, as they would no longer be destroyed instantly through collisions with other par-ticles. So in the early stages of the universe, expansion and cooling down provided the con-ditions for the first build-up of matter. After a tiny fraction of a second, triples of quarks started to form stable protons and neutrons, the building blocks of nuclei. Their binding was caused by the strong nuclear force. After about three minutes, these protons and neu-trons started forming the simplest nuclei such

as deuterium, helium and a tiny bit of lithium. Knowledge of the weak nuclear force allows us to calculate the quantities of the light el-ements formed at this time, as the decay of neutrons was a crucial factor in the process. When the temperature decreased further, all available particle and antiparticle pairs recombined into pure energy. The majority of (anti)matter was converted to radiation at this stage and relatively speaking only a tiny bit of matter was left. After about 300,000 years, the remaining positively charged nuclei started combining with the free electrons to form the simplest electrically neutral atoms, mostly hydrogen (75%) and helium (25%). We have now moved down to the atom at the right-hand side of the Ouroboros' circle. Note that these are still the very early years of the universe, because its present age is 13.7 ± 0.2 billion years. This time, not surprisingly, the electromagnetic force caused the binding. It must have been a dramatic event, because the contents of the universe went through a phase transition from a hot charged plasma to a hot gas of neutral particles. Remarkably, this is when the universe became transparent, as the charges that used to scatter light became bound in neutral atoms, after which light could freely propagate through space.

After the electric force had literally been neutralized, the much weaker gravitational force became dominant. To continue our story we now have to move from the right-hand side of the Ouroboros to the left-hand side. The gas of neutral particles started gravitating towards small inhomogeneities, and those became the seeds of the large-scale structures we see today. After a while the core of those gas clouds became so dense that they started heating up under the enormous gravitational inward pressure, so that finally nuclear ignition occurred spontaneously and stars were formed. Many, very many suns started shining. The universe was filled with trillions of gigantic, hot fusion reactors, furnaces where ultimately the more complicated nuclei were brewed,

such as carbon, nitrogen, phosphorus, and eventually also heavier elements like iron. The atoms that are vital for life were formed this way. Many atoms in our bodies have a rather violent past: they went through the interior of numerous generations of stars that died in violent supernovas, before finally ending up in the debris that formed the earth.

In the formation process of stars it is common for accretion discs to form, from which planetary systems evolve. In all, it took the universe about eight billion years to produce the raw materials for planets like the earth. So now we are approaching the lower left side of the Ouroboros.

A remarkable combination of fortunate coincidences sets the earth apart from the other planets. These include the presence of water and an atmosphere, the stabilization of the orientation of the planet's axis by the moon, and protection from the impact of too many large asteroids by big planetary companions like Jupiter and Saturn. It means that on our blue planet we enjoy a very mild climate with relatively small temperature fluctuations. These circumstances allowed the formation of the large and complex molecules required for life.

Presumably, life formed spontaneously somehow, though I must tell you that there is no hard evidence for this, let alone a detailed understanding of how and where this actually happened. Inanimate matter has never been brought to life in a test tube so far. Yet somehow, complex molecules started to form and combine into organisms that reacted to their environment and passed on their characteristics to their descendants. And this is where we skip over to the right side of the Ouroboros once more, to follow the circle downwards. Life on earth evolved from single-cell organisms to far more complex creatures – and as far as we know, the most complex specimen of all is *homo sapiens*, that is: you and I.

Looking back at those tremendous 13.7 billion years of trial and error, we can only say that the human being is a very curious conglomerate of elementary matter indeed, shaped by the laws of physics, chemistry and biology. This is the wealth of the Ouroboros: she tells us about the building blocks, the structures on different scales, but also about the history of exactly how and when these structures came into being. One sees that we have extended the evolutionary path of life as described by Darwin into the domain of dead matter all the way back to the big bang. It all started out with a huge outburst of matter and energy in its most rudimentary form, growing more and more complex due to the expansion and cooling, until larger structures developed, life broke through and evolution took over – at least in those places where the physical conditions allowed it to.

We have explained the scientific evidence based account of the universe's evolution obeying the general laws of physics and biology from the Big Bang up to the present. Now we should ask what these laws teach us about the future. Looking at this question through the eyes of our scientific theories we arrive at the paradoxical conclusion that in fact the near future is in many ways more uncertain then the very long-term future where we know that certain things will happen for sure. So if we skip the man and/or nature induced catastrophes connected with climate in the short run, we know that on the scale of tens of thousands of years there are various periodic changes in the earth motion and the orientation with respect to the sun that will certainly affect the climate significantly. On scales of tens of millions of years we have to worry about large impacts of asteroids that may occur and destroy life to a large extent, like the one in Yucatan of 60 million years ago which caused the extinction of the dinosaurs. Looking even further ahead we know that in about 5 billion years the sun will stop emitting its vast amounts of energy because it runs out of nuclear fuel after which it will collapse in a supernova forming a giant burning ball extending over a large part of the planetary system. This is an event that will take place with probability one and if by that time we have not invented ways to spread out over larger parts of our galaxy the conclusion will be as grim as inescapable: the tree of life to which all life on earth belongs will be exterminated.

Beyond the scale of stars we have to ask the ultimate question about the future of our universe as a whole. And surprisingly, in recent years we have obtained almost irrefutable evidence for what will happen. To find out we had to make precise measurements of the average energy density, which is very hard because there maybe forms of energy there where we are not very familiar with such as dark matter and as we know now, dark energy. We have done so and the outcome unmistakably says that our universe will keep expanding forever. This means that ultimately all structure will decay in the lightest forms of energy, and the great story ends with an evermore diluted gas of some elementary particles and radiation.

I have presented you with a multilayered view of the world around and inside us, based on the Ouroboros, the symbol of unity in nature. Keeping these layers in mind, you might come to appreciate the thesis of this book: *one nature, one science*. The oneness becomes even more evident if you think of all the cross-links, not only between the layers, but also between the various subfields of science – a tremendous network, of which only a small part has been depicted in the figure.

The arrows indicate cross-fertilization between research fields, but there is more to them than that. The arrow between the nucleus and the human figure, for example, stands for MRI imaging, radio therapy, X-ray imagining and the use of radioactive tracers in the medical sciences. The interconnectedness of all these fields of science also makes the overall scientific worldview very robust. It means that you cannot just walk in and change something on one side, without taking into account what happens at the other end of the arrow. And as there are very many bidirectional arrows in the figure, that makes it hard to come up with an idea that will change the picture significantly. It is hard, but not impossible, and it depends on the abilities and commitment of the scientists involved. Being a scientist is not just a job, it is a way of life, some people say.

I would like to discuss two of the more illuminating connections. Firstly, the two arrows located at the top. This is what I call the *cosmic shortcut* between the physics of the largest distances on the left side and that of the shortest distances on the right, culminating in the big bang itself. And this finally takes us to the explanation of why I did not draw a line instead of a circle, with ever larger scales to the left and decreasing scales to the right, and ourselves happily in the middle. Why the circle of science, and why does the Ouroboros bite its own tail? Well, imagine that we want to learn about what goes

on in the most remote parts of the observable universe, and we aim our telescopes at the farthest things we can see. The signals we receive consist of light, X-rays or microwaves. These all travel at the speed of light, and because this is a finite velocity, they have been on their way for millions or even billions of years by the time they arrive at our observatories. That means that what we see is actually the way the place looked billions of years ago! Exploring the boundaries of space is the same as exploring the boundaries of time!

Now if we combine this fact with the big bang theory, it tells us that if we would look further and further away, eventually we would end up with signals from the big bang itself. But that theory also tells us that the universe has been expanding all along. If we look back in time far enough, we look at the universe at a stage when it was a lot more contracted. Physics tells us that if matter is compressed in an ever-smaller volume, its temperature will rise – the inverse process of the cooling due to expansion. In the remote past, the universe must have been hotter. If we would go back in time, we would enter an era in which there were no living organisms, then an era in which there was no chemistry, no atoms, and eventually not even nuclei. So you see that the big bang is the endpoint of both our searches for the largest and for the smallest scales. And this explains why the circle closes in the big bang.

The two frontiers of fundamental science which at first seemed the furthest apart, lead to one and the same enigma, the origin of our universe. And indeed, discoveries in those two extreme realms of science affect each other. How beautiful and how surprising! Astronomical observations can teach us about the properties of the elementary particles that should exist. In recent times you may think of the discovery of previously unknown forms of energy and matter, called *dark matter* and *dark energy*. Astronomers say they have to be there, but they have not yet been identified in particle physics. And it works the other way around too: particle physics is able to provide detailed descriptions of mechanisms that explain the abundance of certain types of matter in the universe, or the effects of very early cosmic phase transitions such as *inflation*.[21] Indeed there is now a whole field of research called "astro-particle physics." The exact point where two of the great theories of physics confront each other is where the Ouroboros swallows its own tail.

I will mention another splendid example of how different parts of our Ouroboros meet each other. You can see arrows leading down from the two sides to the bottom of the circle, reminding us of the two arrows pointing down indicating the frontier of complexity in an earlier overlay. The human figures in the middle represent the pinnacle of complexity to be found in our part of the universe – and therefore, in my view, the ultimate subject of study for the natural sciences. It is clear that complexity is to be found in both the macroscopic as the microscopic perspective.

This region of the Ouroboros stands for a very extensive frontier of knowledge that confronts the mystery of life in all its variations. On the left we have the classic disciplines like earth sciences, geology, and paleontology, culminating in Darwinian evolutionary biology. On the right we find the complementary molecular approach of molecular and cell biology. The downward arrows are pointing at us, inescapably leading to the ultimate question about understanding human nature and consciousness.

Let us move on to look at the next overlay, where I delve somewhat deeper into two ultimate questions presented by the Ouroboros.

[21] Inflation refers to a brief period in the very early universe when the universe presumably expanded exponentially fast.

Now that we have looked at the Ouroboros from all sides, it appears that science is left with two profound challenges: the question of origin and the challenge of complexity. In a way, these questions are contained in the table of basic questions discussed in Chapter II: the origin question covers the questions of matter and the universe, as well as space and time, and the issue of complexity can be viewed as dealing with life and death, mind and soul, and society. That may serve as an even stronger motivation to think about what these questions mean in science today.

The quest for the origin

So what is the big bang itself? My answer might come as an anticlimax: We don't know. I have told you that at the earliest conceivable instances, the density of energy and the temperature increase were without limit according to the big bang theory. This is simply a solution to Einstein's equations of General Relativity – the very same theory that also predicts another mind-blowing phenomenon, the existence of black holes! If you collect enough matter (or energy) in a small region of space, then a black hole will form. A black hole has a so-called horizon: nothing can escape from the interior of the black hole, i.e. the region within the horizon; the information about objects that fall in would be lost forever. We don't know anything about what happens inside the black hole, except that everything that falls in would disappear. You can see how this causes a dramatic problem within the theory of gravitation: at the very early stages of the cosmic history, our spacetime would become like a foam of black holes because of the extremely large energy density. But what this really means, is that the notion of spacetime would break down at timescales smaller than 10^{-42} seconds and distances smaller than 10^{-35} meters.

We might say that there is a curtain, beyond which our present-day physical theories somehow lose their validity. Questions about

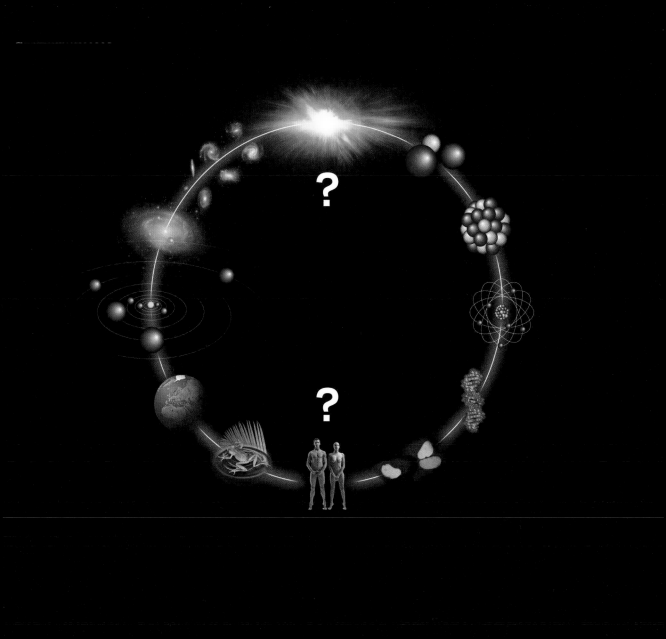

what goes on behind that curtain are unanswerable within the present framework. To answer such questions we need a new overarching framework, bridging the gap between the two pillars of modern physics: the General Theory of Relativity, describing the properties of spacetime, and quantum theory, describing the properties of matter at the most fundamental scales. That is not a modest ambition, and not surprisingly it has attracted many of the brightest minds. They have come up with a brilliant idea called *string theory,* but I should warn you that they are still struggling in the speculative domain. The key idea of this theory is that all fundamental particles, including those carrying the forces – like the photon for electromagnetism and the graviton for gravity – are different manifestations of one universal underlying dynamic entity, namely a tiny string. The different internal vibrational modes of the string correspond to the different particle types, and since the graviton is one of them, the strings necessar-

ily describe the gravitational force.

This is an entirely novel starting point for the description of all of physics. Now you may want to know whether the laws of particle physics and of General Relativity can be shown to follow from this perspective. The answer is certainly yes, but it is only a confirmation in principle, because nobody knows how the standard model of particles exactly fits into string theory. And that makes it very hard to connect string theory with present-day observations. Still, you can probably imagine that such challenges – in which two fundamental theories clash, each of which has survived thousands of confrontations with a variety of experiments – are the most exciting for young ambitious scientists to explore. The crisis can only be overcome by an unusual dose of creative intellectual power, and the result may well be a new turning point, with unexpected consequences for the way we think about nature as a whole.

The challenge of complexity

The theory of evolution taught us to think of increasing complexity as the stages of some dynamical process, which can be governed by surprisingly simple rules. Variation and natural selection are probabilistic in the sense that new developments and changes happen by chance, and yet it is also deterministic, in the sense that it leads to a specific outcome. By modeling and studying specific subsystems in detail, we gain much insight into such processes.

As we now know, the macro-biological process of the evolution of species has a mirror image in the molecular domain; in the structure and evolution of our DNA. DNA is not only some kind of blueprint for life, it is also a beautiful book, or rather an entire library, in which the history of life on earth has been written down in painstaking detail. This is the greatest bestseller of all times, and we have only barely started to read it.

There are many phenomena we would like to learn more about: the origin of life, the origins of language, the understanding of memory and cognition. What are their detailed mechanisms at a cellular and even a basic chemical level? The wide gap between mind and matter has to be bridged. And so we see many new fields emerging, with intriguing names like psycho-chemistry, neuro-psychology, bio-informatics, ecology, population and other types of social dynamics, and astro-biology. They have come out of traditional disciplines, with their own cultural baggage and technical skills, meeting in the quest to understand life in all its astounding manifestations. These new fields have promising research agendas and they are able to grow rapidly because there are many new ways to collect detailed data – think for example of MRI scans of the brain – and powerful computational tools to analyze these data. There is even a certain danger that biology might drown in an overdose of data begging for an explanation.

146

Physics has taught us that the road from experimental facts to a valid theory can be extremely difficult and counterintuitive. Phenomena that at first appear as highly chaotic and erratic may still have a hidden mathematical description that is very simple, and therefore may in the end be more tractable than most of us anticipated. The first steps usually consist of trying to find correlations that are valid over an appreciable domain of the parameter space. These correlations may then be quantified in certain observed scaling laws that may have a broad validity. In the field of complexity, many so-called power laws[22] have been discovered, for example expressing the relationship between the energy-consumption of a living organism and its mass (where the scaling power happens to equal 3/4), or between city size and the number of patents, or for income distributions. The scaling laws that are found form a very compact way to represent large bodies of data, and they may give crucial hints about where possible explanations of the phenomena under study should be looked for. This may be a geometrical constraint or a hint at an underlying fractal structure (reflecting a kind of self-similarity of the phenomenon on different scales), or a particular type of stochastic process at work.

In economics there is a long tradition of sophisticated theory building, but the field used to suffer from the lack of a clear-cut body of reliable data. Lacking hard data researchers used to resort to theoretical assumptions, such as the rationality of consumers or a market equilibrium, which in many cases may not hold. It is hoped that in this digital era, with computers accurately keeping track of certain sub-areas such as the trading on financial

[22] A power law states that a quantity y depends on another quantity x in a way given by the mathematical relation $y = ax^\alpha$, where a is a constant and α the power. Recall that the inverse square laws for the electrical and gravitational forces are power laws with exponent $\alpha = -2$.

markets, or consumer behavior, economics will become more experimentally and observationally driven.

Newspapers are full of discoveries made by many of the new sciences, but these have not yet yielded any really fundamental breakthroughs. My belief is that new turning points will show up, but it may take more time than some of us expected. There are people who say that biology is so fundamentally different from physics that it is naïve, if not plain wrong, to expect similarly strong laws, for example because the field involves many forms of collective behavior. It has always been a matter of looking at seemingly intractable problems from the right angle, with the appropriate tools. After trying long enough, this generates the crucial insight. In fact, the field of chaos theory and nonlinear dynamics can serve as a nice example. When large scale computing became available, one could start investigating nonlinear systems that were impossible to solve analytically, and one learned that even simple systems could develop highly nontrivial behaviour known as deterministic chaos. It became the starting point of a new field of research, with important applications in numerous other fields such as ecology and economics.

Biology concerns complicated networks whose structure and interactions are dependent on the evolutionary history of the organism. Therefore it would be highly improbable to find any universal causal mechanisms. I do not agree. Evolution is a way nature finds solutions to the very generic problems that life or better survival poses.

A cell is a very complicated system indeed, in which very complicated networks of proteins perform innumerable tasks. To figure all this out will take a long time. But science is a lengthy endeavor anyway, and biology is no exception in spite of powerful sequencing machines, MRI devices, and supercomputers.

Let me end this subsection on complexity with a quote from the novel *Saturday* by

148

Ian McEwan, in which the main character – a neurosurgeon – reflects on the future of neuroscience:

For all recent advances, it's still not known how this well-protected one kilogram or so of cells actually encodes information, how it holds experiences, memories, dreams and intentions. He doesn't doubt that in years to come the coding mechanism will be known, though it might not be in his lifetime. Just like the digital codes of replicating life held within DNA, the brain's fundamental secret will be laid open one day. [...] Could it ever be explained how matter becomes conscious? He can't begin to imagine a satisfactory account, but he knows it will come, the secret will be revealed - over decades, as long as the scientists and the institutions remain in place, the explanations will refine themselves into an irrefutable truth about consciousness. [...] The journey will be completed. He is certain of it. That's the only kind of faith he has. There's grandeur in this view of life.

Ian McEwan, 2005

We have come to the end of our kaleidoscopic tour. The Ouroboros, the mythical snake that swallows its own tail, offered us a coherent view of the many levels at which we understand nature and their interconnectedness. We have seen, that in order to understand the right-hand side of the Ouroboros, you have to understand the left-hand side; to understand the bottom, you have to understand the top.

There is only one nature, and as all its parts are connected, that implies that there is also only one science – a science slowly taking shape through the work of thousands of scientists spread out over the earth and in time. It is like putting a gigantic jigsaw puzzle together: at first one starts by just joining separate pieces into small islands, which are subsequently linked together to make bigger chunks, until every piece falls into place and we get a grand view of the whole.

This exemplifies what I like to call the cultural wealth of the natural sciences. In the final chapter I will analyze the remarkable fact that in our daily cultural life, and in particular in the media, this wealth is pretty much marginalized and ignored.

ratione salua manente; nemo enim conuenientiore allegabit
q̄ ut magnitudini orbium multitudo temporis metiatur; ordo spha-
rarum sequitur in hunc modum: a summo capientes initium.
prima et

suprema omnium est stellarum
xarum sphaera seipam
et omnia continens
Ideoq; immobilis
Nempe vni-
uersi locus
ad qu͡e
mot
us
et
p
o

1 Stellarum fixarum sphaera immobilis
2 Saturnus xxx anno reuoluitur
3 Iouis xij anorū reuolutio
4 Martis bima reuolutio
5 Telluris cū luna. an. re
6 Veneris nonimestris
7 Merc xxc dierū
Sol

ſi
·
tio
re-
teorū
omniū
syderum
consideratur
Nam quod
aliquo modo illa
etiam mutari existimat
aliqui: nos alia, cur ita appareat
in deductione motus terrestris assignabimus causam. Sequitur
errantium primus Saturnus: qui xxx anno suum complet circu-
itū post hunc Iupiter duodecennali reuolutione mobilis. Deind
Mars qui biennio circuit. Quartum in ordine annua reuolutio locum optinet: in quo terra cum orbe Lunari tanquam epicyclio
contineri diximus. Quinto loco Venus nono mense reducitur

I can reckon easily enough, Most Holy Father, that as soon as certain people learn that in these books of mine, which I have written about the revolutions of the spheres of the world, I attribute certain motions to the terrestrial globe, they will immediately shout to have me and my opinions hooted off the stage. [...]

I was in great difficulty as to whether I should bring to light my commentaries written to demonstrate the Earth's movements or whether it would not be better to follow the example of the Pythagoreans and certain others who used to hand down the mysteries of their philosophy not in writing but by word of mouth and only to relatives and friends – witness the letter of Lysis to Hipparchus. They however seem to me to have done that not, as some judge, out of jealous unwillingness to communicate their doctrines but in order that things of very great beauty which have been investigated by the loving care of great men should not be scorned by those who find it a bother to expend any great

energy on letters – except on the money-making variety – or who are provoked by the exhortations and examples of others to the liberal study of philosophy but on account of their natural stupidity hold the position among philosophers that drones hold among bees. [...]

Not a few [...] learned and distinguished men [urged] me to refuse no longer – on account of the fear which I felt – to contribute my work to the common utility of those who are really interested in mathematics: they said that the more absurd my teaching about the movement of the Earth now seems to many persons, the more wonder and thanksgiving will it be the object of, when after the publication of my commentaries those same persons see the fog of absurdity dissipated by my luminous demonstrations.

Nicolaus Copernicus
(in De Revolutionibus Orbium Celestium,
in the dedication to Pope Paul III)

Facts speak for themselves

Even in the mythical era, people had high expectations about the magnificent profits to be gained from wisdom and knowledge. Fundamental knowledge would turn us into little gods and provide us with enormous power, if not over nature then at least over our fellow men. Hence it is not surprising to see that the "research" activities in those days had an esoteric and elitist character: they were secret cult-like performances by members of privileged inner circles.

It is a sobering thought for the modern scientist to realize that the roots of science go back to superstition, secrets and plain lies. Yet this somewhat dubious ancestry has led to a most successful offspring in the later natural sciences, based on logical reasoning and experiment. Science opened up a new mental option to cope with the world, the option of not just believing things, but rather critically investigating them. This "free-thinking" attitude often posed a serious threat to the accepted rules and beliefs of established institutions. The confrontations between science and such institutions which took place throughout history were in many ways unavoidable.

In spite of the general public's everlasting interest in spiritual and supernatural phenomena and explanations, most of us will agree that the greatest turning point in the history of consciousness was the discovery that there is a *rational order and structural hierarchy* in our universe. The natural sciences with their demystifying powers started to develop and took the stage with knowledge that really worked. Witches, medicine men and rainmakers began losing ground – but slowly, presumably because knowledge does not imply wisdom. On the other hand, it doesn't exclude it either.

The fact is that science as such does not address moral issues. It does not provide ethical norms, consolation, love, and other existential items which are instrumental to most of us for leading a happy or fulfilling life. From a strictly scientific point of view there is no necessity to engage in such matters, and in fact science is not equipped to do so. In that respect science leaves what many experience as a considerable existential gap.

Presumably that is why religion remains one of the pillars of the social edifices we have erected. Many people are still troubled by old questions like: "Did God create man, or did man create God?" Can science tell us anything useful about such eternal "chicken or egg" questions? Let me first state that any attempt to have a discussion about these questions which deserves scientific credibility is severely frustrated by the fact that the notion of God is one of the most fluid and ill-defined concepts we know of. This is evident from the notable difference between the concept as it appears in different religions. It is used by too many and understood by too few.

I have already discussed some instances of the uneasy relationship between science and religion in Chapter I, and we are not going to dwell on it here. The intrinsic vagueness inherent to most myths, is presumably the main reason for their persistence and apparent vitality. This also applies to revelation-based knowledge. If it is a Miracle, any sort of evidence will do, but if it is a Fact, proof is necessary. Not only is "God" an unbounded, singular concept, it has also undergone innumerable transformations and facelifts throughout history, making it an irrefutable hypothesis. I do not think we should expect that science will ever prove that a God does or does not exist; the best one can hope for is that it helps narrow down what the concept may consistently mean.

If we take into account all the ways people think about God in different parts of society, (s)he becomes a kind of multifaceted diamond, simultaneously being eternity, a superhuman "Big Brother watching you," Love & Peace, the Path, the Soul, and the Teacher, as well as a bunch of mathematical formulae making up a Theory of Everything. On the one hand religion is an effective theory of life, decreeing rules for moral conduct *et cetera*, and on the other hand it is hailed as a last stronghold against the flagrant arbitrariness of the evolutionary process (of which religion presumably is a product itself).

Many scientists believe that we will be able to understand our universe without ever invoking the God concept. They believe that the facts speak for themselves, and they trust that the answers to the hard questions of life will be given by Nature itself. For that reason, scientists will – if possible – refrain from (officially) commenting on religious matters. On the other hand, it has been said that in our materialistic era scientists are the only deeply religious people. I don't know what to make of this, but it is true that many great scientists have commented on the subject matter. It shows that where science itself has nothing to say, scientists may still talk and argue. Just like ordinary people.

Discourse on the method

The greatest minds, as they are capable of the highest excellences, are open likewise to the greatest aberrations; and those who travel very slowly may yet make far greater progress, provided they keep always to the straight road, than those who, while they run, forsake it.

Rene Descartes, 1637
(in Discours de la Méthode)

A goal of science is to describe the largest possible part of reality with the fewest possible variables and falsifiable relations between them. Reality can in my view be described as the total collection of in principle measurable quantities or observables. In this perspective, science is nothing but the identification and systematic reduction of the number of variables and relations. If we think about it that way, science has only just started. To achieve its goals we have to develop and exploit rational analysis and logical arguments (for instance mathematics), and we have to make critical observations to falsify our models and to develop the instruments needed in order to make those observations.

This is in essence the empirical reductionist doctrine of the era of Enlightenment. As I explained in Chapter II, it amounts to a rather pragmatic view which does not require many philosophical nuances. Because variables do not have to be observables themselves – in quantum physics, for instance, they often are not – the interpretation thereof, and in particular the question whether they are a truthful representation of reality, can be left alone for the moment.

Plenty of books have been written about these considerations, but they do not concern us here. What is important, is that even small parts of the scientific enterprise can satisfy the empiricist criteria. Doing good science is in an operational sense conforming to a local set of criteria, that can be applied everywhere and at any instant. It does not depend on whether the researcher believes that an objective or absolute truth is to be found at the end of the journey and what that truth might look like. Though in science we can certainly identify progress, it just is not known whether there will ever be an endpoint, as science as we know it now is an incomplete body of knowledge. Yet the scientific process is a directional one in which, to stay within the context of this book, myths are systematically transformed into robust knowledge.

At this point I should mention the contribution by the philosopher of science Karl Popper. He clarified the issue by putting forward the crucial requirement that a scientific model has to be *falsifiable*; it must be possible for it to be proven wrong, because otherwise it would be useless. Falsifiability refers to the simple fact that it is possible to prove that a hypothesis is absolutely wrong, but not to prove that a hypothesis is absolutely true.

Verifying a hypothesis is only possible in a limited sense, because we can always only check it for a finite number of cases. You might find abundant evidence in support of a hypothesis, but that still does not prove it to be right. A corollary of this statement is the assertion that science is by definition incomplete.

A solid falsification can hit hard and deeply affect the group or individual scientist concerned. Imagine, you have just proven this absolutely gorgeous theorem, and a friendly colleague at the other side of the globe produces a subtle counterexample in a footnote of his next article. Imagine, you have just discovered a new effect – it already bears your name – and a much-respected colleague finds nothing in a carefully executed "identical" experiment.

Falsification may take different forms. It can be done by reproducible experiments whose outcomes contradict the hypothesis, but it can also be achieved by showing incompleteness or inconsistency in the line of argument. Again, there is no such thing as "the correct theory" in absolute terms; we have to live with a "successful" or a "widely accepted" theory. The modest name "standard model" for the most overarching theory of matter at our disposal is an illustration in point.

In view of the falsification criterion, it is better to come up with a theory that is wrong, than with one that is vague. This dictum sounds innocuous, but it can act like an ax in the world of the sciences or would-be sciences. Popper himself, for example, did not consider psychoanalysis and history to be proper sciences. Generally speaking, his ban concerns research fields that only interpret, i.e. in which the interpretation of facts is basically a narrative. A narrative seldom allows for falsifiable predictions, and therefore it has a hard time transcending the speculative stage. In the first chapter I referred to the fact that victors tend to rewrite history, replacing one belief by another. Even so, there must be an overwhelming amount of historical data on the crimes against humanity committed by dictators, suggesting that also in the field of history a fairly reliable comparative theory could be developed, perhaps even with some predictive power.

But within the context of the natural sciences, too, vague theory can be hard to avoid – certainly as a transient state. The first groundbreaking steps towards quantum theory were in fact rather obscure. They were strongly based on *ad hoc* assumptions, leading to a quite unsatisfactory situation which was only resolved much later. Apparently we have to accept the fact that science goes through some "mythical" stage when a dramatic change of paradigm is taking place. *In such instances, science is not so much a matter of finding the right answers as a matter of asking the right questions* (to use a famous quote by Claude Lévi-Strauss). In the next section, we will look briefly at string theory as an up-to-date example of such an emerging paradigm.

A stringy intermezzo

String theory's development as a proposed "theory of everything" has ever since its inception in 1983 been marked by many ups and downs. As mentioned in Chapter III, this theory is an attempt to embrace all fundamental interactions, including gravity: it is the theory meant to bridge the wide gap between General Relativity as a description of spacetime, and quantum theory as a description of fundamental forces and matter. Its basic postulate is that all particle species (observed and yet to be observed) correspond to different internal vibrational modes of extremely tiny superstrings. The idea is that ultimately everything, *matter as well as space and time*, is made of this unique type of string.

The theory immediately exhibited a number of very attractive features: it offered a totally new but consistent quantum description of gravity, and in principle it should be able to accommodate the standard model of particle physics. In establishing the unification of all forces, it would realize researchers' ultimate reductionist dream scenario, describing the whole of nature.

However, to test this theory directly one would have to detect the fundamental strings. And those are very much out of reach, even for the most powerful accelerators that might be built in the not-so-near future. This led to a verdict by some scientists that the theory would never be falsifiable in the strict sense and should therefore be expelled from the scientific arena. Pressure was put on the many theorists working in this field to try hard to find other, more indirect experimental signatures of string theory. Indeed, there are two ingredients of the theory that could be helpful in this respect: (i) the theory is only consistent in ten spacetime dimensions, and (ii) it predicts that nature has a symmetry that has been hidden up to now, called *supersymmetry*, meaning that all particle types would somehow have "superpartners." This is not unlike the prediction of antimatter by Paul Dirac in 1928, only then it was rather easy to prove the existence of antiparticle species experimentally. The superpartners have not (yet) been observed, and nobody knows exactly what their masses should be. The ten-dimensional feature of string theory is also uncomfortable, because nobody has ever sensed any extra dimension. String theory assumes that these dimensions could be very small, so it is not surprising that they have escaped observation so far. Anyway, because of such features many scientists consider the theory as not falsifiable, and therefore according to them it should be disqualified from the great race of science.

In fact, string theory gives rise to quite a separation of minds, with strong proponents like Ed Witten of the Advanced Institute in Princeton and Nobel laureate David Gross, now director of the Kavli Institute for Theoretical Physics in Santa Barbara, and opponents like Nobel laureate Sheldon Glashow and more recently Lee Smolin of the Perimeter Institute near Toronto. He even wrote a whole book about it. A famous quote by Sheldon Glashow and Paul Ginsparg from a 1986 issue of *Physics Today* illustrates the heat of the debate:

Contemplation of superstrings may evolve into an activity as remote from conventional particle physics as particle physics is from chemistry, to be conducted at schools of divinity by future equivalents of medieval theologians. For the first time since the Dark Ages, we can see how our noble search may end, with faith replacing science once again. Superstring sentiments eerily recall "arguments from design" for the existence of a supreme being. Was it only in jest that a leading string theorist suggested that "superstrings may prove as successful as God, Who has after all lasted for millennia and is still invoked in some quarters as a Theory of Nature"?

Twenty-five years have passed and the situation is still not settled. Of course there are hopes that with the Large Hadron Collider at CERN a superpartner will be caught or maybe that some energy will be seen to be leaking away into the extra dimensions. That would be tremendously spectacular, for sure, but for the moment it is sheer hope.

And what if nothing stringy shows up? There are secondary aspects to be taken into account. The theory lacks predictive power, or rather its predictions cannot be tested, but mathematically it is very rigid and restrictive. In string theory physics and mathematics have found a productive arena for profound exchanges. For some results, Fields medals (the mathematical equivalent of the Nobel Prize) have been awarded. Innovative ideas and methods with elusive names like *topological field theory* and *conformal field theory*, which were mostly developed in the early days of string theory, have now found useful applications in other fields of physics and mathematics.

In conclusion one could say that string theory is barely tolerated by the hard-core empiricist, but it still gets the benefit of the doubt because of its intriguing creative relationship with mathematics. Is this unprecedented? I am not so sure. There have been instances before in which physicists resorted to the free creative principles of mathematics in order to make progress. Two instances come to mind. The first is the 1931 article by Paul Dirac, in which he predicts the existence of magnetic monopoles, i.e. particles that carry an isolated north or south magnetic charge. This was revolutionary because all magnetic phenomena observed in nature could be understood from magnetic dipoles (say, tiny bar magnets) only. In the introduction to this beautiful but speculative article, Dirac makes the following remark:

Actually the modern physical developments have required a mathematics that continually shifts its foundations and gets more abstract. Non-euclidean geometry and noncommutative algebra, which were at one time considered to be purely fictions of the mind and pastimes for logical thinkers, have now been found to be very necessary for the description of general facts of the physical world.

To continue with:

Quite likely these changes will be so great that it will be beyond the power of human intelligence to get the necessary new ideas by direct attempts to formulate the experimental data in mathematical terms. The theoretical worker in the future will therefore have to proceed in a more indirect way. The most powerful method of advance that can be suggested at present is to employ all resources of pure mathematics in attempts to perfect and generalize the mathematical formalism that forms the existing basis of theoretical physics, and after each success in this direction, to try to interpret the new mathematical features in terms of physical entities.

Einstein too showed some mildness in this respect. He was clearly in favor of some room for free explorations in the domain of theoretical and mathematical physics, which may offer the string theorists some consolation in their profound struggles. In 1934 he wrote:

The theoretical scientist is compelled in an increasing degree to be guided by purely mathematical, formal considerations [...]. The theorist who undertakes such a labor should not be carped at as "fanciful"; on the contrary, he should be granted the right to give free rein to his fancy, for there is no other way to the goal.

So there are notable achievements, but the Holy Grail of a "theory of everything" is still up in the air, and may stay there for some time to come. String theory's limited success in meeting today's observations should not surprise you, because the action in string theory typically takes place at an energy scale some 10 to 15 orders of magnitude higher than what is presently accessible in accelerators! It seems like string theory, with its firm but not-falsifiable predictions, will be with us for some time to come – an emerging paradigm which, if confirmed, would be a true turning point.

Leaving Popper behind?

By and large, in the natural sciences there is a cumulative growth of knowledge which survives the falsification principle. However, if you take a step back and look at these sciences from afar, there may be some reasons for concern. In *The structure of scientific revolutions*, Thomas Kuhn investigates how scientific revolutions take place. Let me paraphrase Kuhn's story, or better: give you a version of how I see this. We start out with a situation in which there exists a governing paradigm on which the scientific community agrees: there is a broad consensus on what is "true" and what is "false." Science is in a state of equilibrium, in which everybody is happy to work within the framework of the established dogma; "normal" science consists of reestablishing the established, "puzzle solving" and filling in the details. Everything seems fine and under control. But alas, at a certain point observations are made that do not appear to conform to the prevailing dogma. Evidence starts piling up that causes cracks in the holy paradigm. Calculations become inconsistent, observations depart from common expectations. The scientific community expends energy to adapt the model and develops repair strategies: competitors critically repeat experiments, and complementary tests are put into place. The most ardent defenders of the prevalent paradigm begin to get nervous. Very slowly, some kind of ominous feeling starts spreading among the members of the community; they realize that they can no longer just repair their theory. Little interventions won't do, because, something is seriously wrong! The situation starts to escalate and to look like an acute crisis, a deadlock; alternatively one might describe it as a collective pregnancy, or the silence before a storm. The sky looks dark and heavy, but nobody knows as yet when and where the lightning bolt will hit. It starts raining new ideas, and everybody is waiting for the creative moment of a breakthrough; a liberating move for a whole generation of researchers. It is a watershed, a phase transition as the body of knowledge attempts to settle into a new ground state – a very unfamiliar state, governed by entirely different rules and concepts which will only emerge after a sufficiently long time. It may not be a true revolution, but it is a turning point that does not proceed along a well-planned, smooth path.

Sometimes it is hard to draw the distinction between revolutionary and evolutionary changes, but many people do view drastic turning points in our thinking, such as Relativity and quantum theory, as revolutions. I already mentioned before that I find the term "revolution" misleading, because it is strongly associated with destruction. The ruling class is put under the guillotine, massacred in their golden palaces; the suppressed express their anger by destroying all symbols of the overthrown regime – an iconoclasm, an expropriation of the Winter Palace – in order to eradicate all traces of the evil past. But that is not what happens in science, which is why I prefer the term turning point or Kuhn's own "paradigm shift." For example, after the "revolution" of Relativity, classical mechanics kept its prominent position, if not in the most fundamental sense, then certainly in the practical sense. Einstein talked with the greatest admiration about Newton, who himself used the maxim: "If I have seen further it is only by standing on the shoulders of giants."

I like to view a scientific turning point as the adding of a new conceptual dimension, the creation of a larger space in which contradictions can be resolved, and facts can be liberated from their fixed and paradoxical interpretation. Whereas we initially were confined to a plane, now we can hover above it and get unprecedented views on the landscape of reality. We may be able to see both sides of a coin at the same time. The notion of a turning point does justice to the *cumulative nature* of scientific progress.

Kuhn actually went much further: he concluded that the worlds before and after the revolution are incomparable, because the conceptual framework has changed dramatically. This was indeed the case with the turning points I have discussed. But if you cannot compare those worlds, it is difficult to claim that you have made progress.

> *We may, to be more precise, have to relinquish the notion, explicit or implicit, that changes of paradigm carry scientists and those who learn from them closer and closer to the truth.*
>
> *Thomas Kuhn, 1962*

Essentially, doubt is cast upon the cumulative growth of science. Popper may be right on a small scale, but according to Kuhn science loses its orientation when it goes through a shift of paradigm. Does that mean that the content of scientific theories and paradigms is arbitrary, because they are the outcome of a rather arbitrary social and historical process? No, this conclusion is not logical and not well-founded. Steven Weinberg, in the chapter entitled "Against philosophy" of his *Dreams of a final theory* makes an amusing comparison:

> *A party of mountain climbers may argue over the best path to the peak, and these arguments may be conditioned by the history and social structure of the expedition, but in the end either they find a good path to the peak or they do not, and when they get there they know it.*

Nobody will conclude that the Mount Everest is an intersubjective social construct, and neither is the formula $E = mc^2$. The fact that there are many roads to a certain truth does not exclude its having a more objective meaning.

The essential point in this discussion is that the worlds before and after the turning point *are* comparable, even if they are not commensurate in the strict sense. Usually the new theory embraces the old theory as a special case, often in a very precise sense as a particular limit. The new theory has to preserve or reproduce what was good about the old one. Newtonian physics is a well-defined approximation of relativity; quantum theory explains classical physics, but definitely not the other way around. And that is exactly what progress means.

Now Weinberg's statement is not a sophisticated philosophical argument, but rather a suggestive analogy. It underscores the sad truth that science and philosophy have disconnected from each other. It was the science philosopher Paul Feyerabend who pointed out long ago that the practicing scientist doesn't need to consult a philosopher about whether a theory should be accepted or not. On the other hand Feyerabend criticized the assumption that theory and experiment would be independent, which by itself turns an objective experimental verification of the theory into a fiction. The validation

process in science depends on the context, and the study of that context as a cultural, historical process should be high on the agenda.

Larry Laudan and Imre Lakatos go even one step further. They distinguish an internal and an external history of science, where the external history cannot be reconstructed in a rational sense, because political and social factors play a decisive role and it involves numerous subjective value judgments. Now this looks to me like a valuable distinction. Indeed we all know that funding policies are real policies, that do indeed depend strongly on the political climate. Whether global warming, terrorist threats, or a megalomanic weapon development program appear at the top of the political agenda strongly depends on the preferences of the ruling parties.

Here we enter the dark side of progress, where many short-range political or economic interests get mixed up with the prioritization of scientific objectives. It is of the greatest importance that the scientific community remain critical and truthful to their highest principles of research, to ensure that long-range truly fundamental research objectives are not strangled in fashionable, trendy hype-type science, or worse, in science that is overtly aimed at extending our destructive powers. History shows that the fundamental attitudes toward science can vary significantly with leadership philosophies. In a 2004 cover article in *The New York Times Magazine*, Ron Suskind reported on a conversation he had with a senior aide of then president G.W. Bush (presumably Karl Rove). In that conversation a rather striking view on reality was expressed, which in retrospect probably perfectly characterizes the regime's strong sense of superiority and misplaced confidence in political and military dominance:

The aide said that guys like me were "in what we call the reality-based community," which he defined as people who "believe that solutions emerge from your judicious study of discernible reality." I nodded and murmured something about enlightenment principles and empiricism. He cut me off. "That's not the way the world really works anymore," he continued. "We're an empire now, and when we act, we create our own reality. And while you're studying that reality – judiciously, as you will – we'll act again, creating other new realities, which you can study too, and that's how things will sort out. We're history's actors ... and you, all of you, will be left to just study what we do."

In his 2008 book on the financial crisis, hedge-fund pioneer and philanthropist George Soros makes an interesting statement about such arrogance of power:

The public is now awakening, as if from a bad dream. What can it learn from the experience? That reality is a hard taskmaster, and we manipulate it at our peril: The consequences of our actions are liable to diverge from our expectations. However powerful we are, we cannot impose our will on the world: we need to understand the way the world works. Perfect knowledge is not within our reach, but we must come as close to it as we can. Reality is a moving target, yet we need to pursue it. In short, understanding reality ought to take precedence over manipulating it.

And he is right: we should look at the excesses that took place in an era preceding the financial crisis, when greediness at the top became generally acceptable and even in vogue. It spread like an infection and brought down the world economy, simultaneously generating a financial and a moral crisis. This was achieved under the hallmark of the free market, which not surprisingly turned out to be dominated by an unbalanced short-term perspective of ever-increasing profits and shareholder value. After the Bush administration completed its term, Barack Obama, in his first big speech on science at the 2009 annual meeting of the National Academy of Sciences, clearly expressed a criticism echoing the same astonishment.

And we have watched as scientific integrity has been undermined and scientific research politicized in an effort to advance predetermined ideological agendas.

I believe that the discourse on method is not on scientists' mind in any practical sense. The problem is, as Gerald Holton already pointed out in 1984 in an article in *The Times*

The perception of a majority of scientists, right or wrong, is that the messages of more recent philosophers, who themselves were not active scientists, are essentially impotent in use, and therefore may be safely neglected.

You may be surprised or even irritated by the cavalier way in which I deal with such serious philosophical matters as the "scientific method." I am sure that I do so because I am a scientific practitioner; over the years I have become tired of an overdose of philosophical, epistemological, ontological as well as strategic and political deliberations. I am however fully aware of the fact that many students love to discuss the philosophical and ethical aspects of science, and often even prefer it over the much harder part of understanding its content. I think they should do so, because it helps them become conscious practitioners of science, and there is a need for such attitude, seeing how important the role of science has become in society. And anyhow, it is a choice that they should make for themselves. The reason for my own above-mentioned fatigue is that I found that such considerations did not add very much to the content of science, and therefore could not give me comparable satisfaction.

Talking with meta-scientists such as philosophers felt a bit like talking to doctors who had already agreed on the treatment, while I as the patient had not even been asked yet to describe my problem or symptoms. Probably many practicing scientists never got beyond Popper's analysis, because it is very recognizable from their own experience. His theory offers them enough room to be productive and creative researchers. The problem is that the philosophy of science has left only very modest traces in the solid body of scientific knowledge. So I would say that the lack of enthusiasm of scientists to start a discussion over and over again about the foundations of their scientific methods should probably not be interpreted as arrogance, but rather as a lack of pretension. They might subscribe more easily to the statement by the French molecular-biology pioneer and Nobel laureate Jacques Monod:

The ethic of knowledge is the commitment to the scientific exploration of nature.

A sociology of science

The fact is that scientists themselves rarely have doubts about what problems should be tackled and about whether their attempts are successful or not. In that sense, peer review and peer pressure are effective instruments to get the priorities right. I think that Kuhn's generalizing conclusions were premature, but even so, his work became the cradle of an intellectual movement that wholeheartedly applied itself to the demystification of science. This movement ended up in the ultra-relativistic attitude of postmodernism and deconstructivism, which has managed to establish itself even at first-class universities. Some of its exponents are so remote from what I think science is about, that I can only consider their striving for a post-rational epistemology – in which science gets reduced to the outcome of a political negotiation between the scientists with power – as being doomed. Only a painful addition of semi-truths can lead to such fallacious statements as that $E = mc^2$ is little more than a slogan in an election campaign! The picture that emerges here is that the postmodernists, because of their single-minded focus on structure and methodology of science, have become totally disconnected from its contents. I will therefore not dwell any further on these developments here, except for one aspect which should trouble us.

As long as they can do their work, scientists may not be bothered very much by what is claimed about their endeavor by a post-rational elite. But the situation may change when these antiscientific forces penetrate government circles or funding agencies, potentially leading to highly undesirable boundary conditions on the practice of scientific research. What I mean is that if the social relevance and implications, or political fashions, become a dominant criterion in setting the agenda for science, its intellectual integrity can be jeopardized, because then we legitimate a situation in which political dogmas are imposed as a pre-scientific choice.

Such ideas about the control of science have deep roots, and they have quite influential advocates at both ends of the political spectrum. The liberal intellectuals have always flirted with these ideas. But also the power structures behind the biomedical, pharmaceutical and military industrial complexes have deeply penetrated in scientific practice. If we go back in time a little, we may quote Theodore Roszak's *The making of a counter culture* in which, based on Kuhn's work, he makes a clear accusation:

The sciences, in their relentless pursuit of objectivity, raise alienation to its apotheosis as our only means of achieving a valid relationship to reality. Objective consciousness is alienated life promoted to its most honorific status as the scientific method. Under its auspices we subordinate nature to our command only by estranging ourselves from more and more of what we experience, until the reality about which objectivity tells us so much finally becomes a universe of congealed alienation. It is totally within our power ... and it is a worthless possession. For "what does it profit a man that he should gain the whole world, but lose his soul?"

The progress of expertise is a "bewilderingly perverse effort":

Consider the strange compulsion our biologists have to synthesize life in a test-tube – and the seriousness with which the project is taken. Every dumb beast of the earth knows without thinking once about it how to create life: it does so by seeking delight where it shines most brightly. But, the biologist argues, once we have done it in a laboratory, then we shall really know what it is all about. Then we shall be able to improve upon it!

Another influential thinker of the progressive movement was Herbert Marcuse, who in his book *One-Dimensional Man* introduced the notion of "repressive tolerance" and talked about artificially created needs. Science is compromised; it is stripped from its autonomy and integrity, and plainly identified as the subdued agent of the established power structures:

The scientific method which led to the ever-more-effective domination of nature thus came to provide the pure concepts as well as the instrumentalities for the ever-more-effective domination of man by man through the domination of nature. Theoretical reason, remaining pure and neutral, entered into the service of practical reason. The merger proved beneficial to both. Today, domination perpetuates and extends itself not only through technology but as technology, and the latter provides the great legitimation of the expanding political power, which absorbs all spheres of culture.

These expressions of fear echo through all times, including the present, and emphasize the alienation process that science and technology have given rise to. It is as if science is the prime enemy of culture and spirituality, as it destroys the poetic dimensions of reality and must therefore be immoral.

Is such criticism outdated? Yes and no. I would say that this anti-scientific message finds a new voice in practically any era, and is usually linked to people with a high moral prestige. Let me close with some quotes from Vaclav Havel, author, poet, and former president of the Czech republic. In an article entitled "The End of the Modern Era" in *The New York Times* in 1992, he basically interprets the excesses of communism as products of the scientific attitude.

The end of Communism is a serious warning to all mankind. It is a signal that the era of arrogant, absolutist reason is drawing to a close and that it is high time to draw conclusions from that fact. [...] We all know that civilization is in danger. The population explosion and the greenhouse effect, holes in the ozone layer and AIDS, the threat of nuclear terrorism, [...] all these, combined with a thousand other factors, represent a general threat to mankind. [...] Traditional science, with its usual coolness, can describe the different ways we might destroy ourselves, but it cannot offer us truly effective and practicable instructions on how to avert them. There is too much to know; the information is muddled or poorly organized; these processes can no longer be fully grasped and understood, let alone contained or halted.

My view is that the excesses and horrors of the communistic era have as little to do with scientific attitudes or reason, as with socialism or democracy. They have everything to do with the ruthless and irrational exploitation by a totalitarian power structure, which is the opposite of the democratic rationalism of science. Havel continues with an appeal to the world:

We are looking for new scientific recipes, new ideologies, new control systems, new institutions, new instruments to eliminate the dreadful consequences of our previous recipes, ideologies, control systems, institutions and instruments. We treat the fatal consequences of technology as though they were a technical defect that could be remedied by technology alone. We are looking for an objective way out of the crisis of objectivism.

These are strong statements, allegations directed at science loaded with grand emotions. They may be alarming, but they are not very clear on what we might want to keep and what we should leave behind. Still, the tone leaves little doubt about the underlying feeling. Many people appear to be afraid that scientific development is an uncontrollable process leading to an apocalypse. But science is our great help and guide, as long as its democratic procedures are well protected, so it can remain an open quest for knowledge. We have to be watchful to make sure it will not be brought down by the quest for power or money.

My personal hope stems from the long-term perspective. I see most political and economical doctrines falling apart or collapsing, while I see science steadily going forward. Its remarkable past, though deeply interwoven with human existence in general, has in my opinion been remarkably insensitive to the particulars of world history. We have seen that, to a large extent, science generates its own questions about nature. Researching those has led to our turning points. These show up in an almost predetermined order, i.e. discoveries are not so much determined by what we want to know, as by what the hierarchy of structure in nature allows us to discover.

Science going Google?

It is important to reflect on those serious concerns about the role of science and technology, which even culminated in the term "the crisis of objectivism." Such concerns breed anxiety in the public opinion. The general public typically does not distinguish between science and technology, but still the overall emotional reaction against science is something to be worried about.

It proves at the very least that the scientists are doing a poor job informing the public about their mission and their results. The scientific community has to come out and make concrete statements to the public about these issues, and to provide answers if possible. We should be active in showing that we share many of the concerns about how this world is managing its problems, in particular on a global scale. The way the "inconvenient truth" about global climate change has managed to penetrate from the scientific community to society as a whole is an example that gives rise to some hope.

One conceivable answer to this supposed "crisis of objectivism" has come from a rather surprising angle: from information scientists announcing the "Petabyte Age." In this era we will have computers at our disposal capable of efficiently dealing with petabytes[23] of data – millions of billions of bytes. They will enable us to mount an unprecedented attack on reality, not by modeling it and using sophisticated mathematical abstractions, but rather by directly investigating it through brute force. There will basically be unlimited amounts of stored data that can be effectively accessed and processed by networks of computers: a gigantic amount of number crunching. The claim is that as they will preferably deal with real data only, simulation or modeling will become obsolete.

Imagine what Google would do if you would ask it to seek for the truth. Will there still be a need for traditional science in the Petabyte Age? Or will PetaGoogle generate a full-fledged replacement for the would-be theory of everything? Petabyte thinking is the ultimate instance of the notion "more is different." Peter Norvig, Google's research director, coined the dictum: "All models are wrong, and increasingly you can succeed without them."

The new type of investigation will revolve around very extensive statistical analysis of huge amounts of data, searching for correlations. And according to the peta-adepts, correlations are enough. They are saying there will be no need for causal relations or conceptual understanding, because reading the massive amounts of data careful enough allows us to make reliable predictions even without understanding the underlying mechanisms. From imaging and measuring the body, the treatment will follow: no need for a doctor, who after all only has a limited knowledge of all cases that have preceded this patient. In the new age, diagnostics are done by correlating. Justice: by correlating. Predicting votes, reactions to political regulations, economic crises, *et cetera*: all based on mere correlations of data. Petamachines will speak all languages, do advanced mathematics, and write letters. In short: *everything you can do, they can do better*. And futuristic though that may sound, sure enough, the first steps in these directions have been taken.

[23] Peta- is a prefix like kilo- and giga-, indicating a very large quantity, namely 10^{15}.

A good thing about the Petabyte Age idea is that it suggests the possibility of maximal transparency and equal access, like the Internet has already demonstrated to provide in many instances. It could be an impressive movement towards an online democracy, in which society is polled continuously and reliably. The whole of society is turned into a computer lab where the human condition is studied – but the big question is: Who is going to see the results of all these polls first? Are they for sale? As soon as highly relevant data on the public domain stay outside that public domain and are turned into commercial "products," there is every reason to worry. Then we may end up with the opposite of what we hope for, namely quite severe manipulation and obstruction of the democratic processes.

Chris Anderson, editor in chief of the Internet magazine *Wired*, summarizes the Peta Gospel as follows:

> *But the opportunity is great: The new availability of huge amounts of data, along with the statistical tools to crunch these numbers, offers a whole new way of understanding the world. Correlation supersedes causation, and science can advance even without coherent models, unified theories, or really any mechanistic explanation at all. There's no reason to cling to our old ways. It's time to ask: What can science learn from Google?*

In Chapter III, when discussing the challenge of complexity, I already alluded to the great progress that is being made particularly in the medical and social sciences due to the availability of large-scale computer power. However, it is still hooked up to the good old "knowledge machine" discussed in Chapter II, in which many other devices such as accelerators, MRI scanners, and space telescopes are instrumental to progress in science. These devices could be built because of our understanding of the underlying principles, by exploiting our scientific and technological expertise. Let us not forget that the computer itself is also a product of that double helix of science and technology. My point is that large-scale computers will be instrumental in making important progress in science, but they will by no means *replace* science. They still need input from observations and measurements, and they need us to ask the right questions.

The beauty and uniqueness of the scientific edifice is that the turning points we have been discussing represent universal insights and laws of nature – and indeed, a single formula is the condensed statement of an infinity of correlations. I could go even further and claim that the laws of nature eliminate the need for an infinite number of measurements. They are more than just an ultimate form of reduction, of data compression.

Analytic methods can sometimes reduce problems which at first sight would need humongous computer power, to quite manageable calculations. Still, there also exist problems for which no algorithm is known to solve them efficiently. An example is the problem of finding prime factors of very large numbers.[24] Once we have the answers,

checking them does not take long: we only need to do a few multiplications. Finding the answers, on the contrary, is extremely hard. Such problems would baffle the petabyte machine. In these cases, apparently bigger is just not different enough, and we might have to resort to something more drastic like *quantum computers*. However, *if* those are ever going to exist, they will not come out of petabyte computing power: they will be the product of the double helix of our more conventional knowledge generator.

Science is a very dynamic process. The resistance in accepting new proposals is proportional to their expected impact, and that is as it should be. In 1993, the British mathematician Andrew Wiles presented a proof of "Fermat's last theorem." This theorem, which is over three centuries old, is as simple to formulate as it is hard to prove: the equation $a^n + b^n = c^n$ with a, b, c and n being positive integers, has no positive solutions for $n > 2$. After submitting his paper, Wiles received quite a few requests from highly esteemed colleagues who acted as referees for the paper, to provide further details on certain steps in the proof. This was not much

[24] Prime numbers are numbers which cannot be divided by any whole numbers, except 1 and the number itself, such as 7, 11, and 59. For non-prime numbers, their "prime factors" are the prime numbers by which they can be divided. For instance, the number 15 has prime factors 3 and 5.

174

of a problem, until a group from MIT stumbled on a fatal incompleteness or circular reasoning in Wiles' chain of arguments. It took him almost two years, in which he came on the verge of giving up the project, to discover the missing link. And where did he find it? In his own wastebasket, in a notebook with pieces he had discarded in an earlier attempt to find the proof! This problem had kept generations of geniuses occupied[25] and much brilliant mathematics had been invented in attempts to prove it, but none of the methods seemed powerful enough to complete the proof until Wiles came along. The new methods which were triggered by the theorem of course found applications elsewhere.

In such matters Google is not much of a help, I am afraid, except that it might come in handy for tracking down the history of the proof.

[25] The interest in Fermat's last theorem has always been somewhat disproportional because of the intriguing fact that Fermat, in a note in the margin of his copy of Diophantus' Arithmetica, not only proposed the theorem but also claimed to have a "truly wonderful proof" of it. Unfortunately, he added: "hanc marginis exiguitas non caperet" or "which this margin is too narrow to contain."

Encyclopedic perfection is not on a par with learning how to ask the right question. So whereas large-scale computing will change our lives in many important and relevant ways, it will not eliminate the more traditional forms of science, and human involvement in them. Google is groundbreaking in terms of sharing and disseminating knowledge, and also for data mining in general. It offers new ways for the bottom-up dissemination of democratic principles and practices. We can already see that governments feel this to be a threat and start impinging on it. But for the generation of new truly fundamental knowledge, search engines and their more sophisticated offspring may at best play an enabling role. I do not share Peter Norvig's Peta-optimism. On the contrary, the unique blend of intellectual and creative power of the human brain is of increasing importance, and we need to develop and exploit it to keep moving forward, not only in our quest for knowledge but even more so in our quest for understanding.

Science at the heart of culture

The feeling persists that no one can simultaneously be a respectable writer and understand how a refrigerator works, just as no gentleman wears a brown suit in the city. Colleges may be to blame. English majors are encouraged, I know, to hate chemistry and physics, and to be proud because they are not dull and creepy and humorless and war-oriented like the engineers across the quad.

Kurt Vonnegut

Science has a cultural dimension. This is unmistakably evident from the philosophical and theological shockwaves caused by its turning points. The Earth turned out not to be flat, wasn't located in the center of the solar system; far worse, science has deported us humans to a rather arbitrary outskirt of the universe, maybe even of *a* universe in the outskirts of *some multiverse*. The relativity of space and time has made the notion of a dynamical universe virtually inescapable, leading to important insights into cosmic evolution and the narrow time window for life on Earth. Again our claim to uniqueness had to be toned down: we have to open our minds to the possibility that life might well have developed in many other places. The observation of many exoplanets orbiting faraway stars and the relative complexity of the chemistry in stellar clouds may turn out to be a first step towards discovery of life elsewhere.

In the microcosm, there is the profound uncertainty that showed up at the most fundamental level of nature through the discovery of quantum mechanics. And also our intimate genetic relationship with the primates as our nearest neighbor on the evolutionary ladder, a message that for many is hard to swallow – again our place in the universe turned out to be less prominent than we had always thought. And what about the molecular basis of all life, which suggests that many aspects of life, even of our emotional life, are just chemistry? Perhaps next week in this theatre: the chemistry of the brain.

The fact is that the turning points of science cause major changes of perspective outside science as well. Obviously they have an impact on society. Yet they fit badly into the realms of established culture; there is a painful lack of appreciation for the hard sciences in our cultural life.

A similar conclusion can be drawn if we focus on the ever-increasing impact of technology. Sometimes the social dimension of the natural sciences becomes most painfully visible in our daily life. It started with tools and means of transport, allowing us to trade and to explore the world. Questions about the maximally possible efficiency of steam engines generated the field of thermodynamics, and from that emerged the idea that different forms of energy could be converted into each other – a notion which in turn opened the door for the construction of machines powered by gas, oil, and later electricity and even nuclear energy. This early mechanization culminated in the industrial revolution.

Then, some 100 years later, after the introduction of the computer, we enter the era of automation of production processes, culminating in today's massive employment of industrial robots. And now we witness the next step, a consequence of the ever-larger scale of integration of electronic circuitry, processors and memory chips, as we have entered a global information age. The software running on machines is able to perform the most incredible tasks with an incredible accuracy. This is information processing in its most abstract sense. Endless strings of zeros and ones represent mail or chat, salary administrations, clumsy holiday pictures or brilliant calculations. To the computer, information is information, and it is willing to perform any operation on it, store it away in any desired format. This processing boom has led to a hierarchy of formal languages stacked on top of each other, allowing us to talk with computers and allowing them to talk to each other about what to do with all those zeros and ones. A rich world of artificial linguistics has emerged.

And at present we are also witnesses to rapidly developing insights in molecular biology, which will have ever-more direct consequences for our physical and mental wellbeing. This technological development has only barely started.

It is evident that scientific knowledge and technology have been at the root of unprecedented social revolutions. I cannot think of any political, religious or economical doctrine that has been able to bring about such radical and robust changes, nor any world leaders. Sure, these great leaders may have facilitated the pursuit of scientific and technological developments in varying degrees – or, for that matter, hindered them – but in the end the inventions took place in the world of science, the world I have described in the chapters above.

In fact, society has always been pleased as well as displeased with the possibilities that emerged out of our growing insights in the working of nature. In my opinion, this dichotomy occurs because the societal awareness of what is going on in the basic sciences is so severely lagging behind, or even plainly absent. "It is possible to live and not know," Richard Feynman once quipped. Yet a 2009 survey of the American Association for the Advancement of Science, the world's largest scientific organization, indicates that at least two-thirds of the American public hold scientists in high regard. The feeling is hardly mutual. The report also said that 85% of the scientists are of the opinion that the public ignorance of science poses a serious problem. And many of them also rank the public coverage of science in papers and other media as poor.

The discovery of the structure of DNA may serve as an example. It took place in a shed on the inner court of the old Cavendish laboratory in Cambridge, in the United Kingdom, in 1953. It was a turning point which, as we now know, has deeply influenced our way of looking at life – whether it concerns illness and health, medicine, our talents, or manmade clones. Extra bonuses are that we can now convict criminals for crimes they committed long ago, and also prove the innocence of convicts who may have been buried a century ago! Interestingly, a considerable societal reaction to the new opportunities raised by the knowledge of DNA only came after about fifty years, after a mouse with an ear on its back appeared in the news and the cloned sheep Dolly was shown on tv.

I like to emphasize once more that the turning points in the sciences have been the dominant driving force behind many social innovations that have led to the emancipation of workers, women, minorities and now of children. Science doesn't care about barriers between classes or sexes, or about ethnicity. Scientific knowledge is a unique example of a shared global commodity, and in the Internet age it is spreading at an unprecedented rate.

These developments underscore the importance of ensuring that in all branches of education the basics of science and mathematics are taught. It would be great if the average citizen would be more conversant on scientific and technological matters, exactly because there is no guarantee that the applications of science will become a blessing to society. We can only try to make them so, if we are equipped to take part in the debate. We need education to enable us to have a reasonably independent and well-founded view on the questions that lie ahead of us; and that encompasses more than marching in a protest demonstration – but it does not exclude that either. The fact is of course that new and basically innocuous knowledge may well be compromised along the way to its application. Our search will not stop, nor should it, exactly because it has made many lasting improvements to our living conditions. The question is not whether we should stop moving ahead, but to ensure that the implementation in society is to our benefit.

Atoms, energy or DNA are neither good nor bad, they are neutral in that they merely reflect facts of nature. The following quote is attributed to four-term US senator Patrick Moynihan: "We are each entitled to our own opinion, but no one is entitled to his own facts." And knowledge of the facts represents a tremendous potential. This knowledge is indispensable if one has to decide on whether or not to implement this or that specific application. For example, the decision to invest on a very large scale in bio-fuels may or may not be driven by the wrong insufficient factual considerations, and in the end may turn out to be counterproductive.

Sometimes scientists take the lead in informing the public of problematic or threatening perspectives, like in well-known cases of nuclear safety and climate change. Society has to deal with these problems and to create scenarios for their solution. I certainly would not like to let these problems be 'solved' by market forces alone, because the conflicts of interest may lead to strategies that in the end do not solve our problems at all. These aspects clearly need the full attention of our societal institutions, and there needs to be debate not only in the dark backrooms of those institutions, but preferably up front in the public arena.

I am convinced that genetic engineering will one day enjoy the same kind of innocent status as, for example, the small operative interventions we do today for medical or even æsthetical reasons to correct lips, noses, buttocks, breasts and bellies, in addition to the world of dental rectifications. For that reason, social, political and educational organizations should contribute to this human evolutionary process in an appropriate way. But then these institutions should know what they are talking about, so they can take decisions when needed, and recognize what they want and don't want. This I would call an advanced civilization.

Sadly, we have the phenomenon that it is perfectly acceptable to say that you "don't understand anything" about science and technology. Easy ignorance. I have always found this highly suspicious: are these people unable to or do they simply not want to know, I wonder. This painful misalignment between prevailing culture and the natural sciences echoes the title of the famous lecture *The two cultures* held by C.P. Snow, in 1959. In it, he remarked that if you would ask a well-educated person to talk about the second law of thermodynamics, usually a painful silence would follow. "Yet, I was asking something which is about the scientific equivalent of: *Have you read a work of Shakespeare's?*" I am not convinced that all that much has changed since 1959. Fear and a kind of suspicion exist as it comes to a scientific discourse or debate. A well-known author and essayist confided to me with a smile: "Math is something that I leave to my personnel." So I answered: "Great, and where did you find such personnel?"

Scientific illiteracy is widespread under intellectuals. Nobel Prize winning physicist Sheldon Glashow – who introduced the Ouroboros of Science – stated in the foreword of his book *From Alchemy to Quarks*, in a somewhat grumpy manner:

Some people regard scientists as cultural illiterates, unable to write, unwilling to read, captives of their narrow expertise and deserving candidates of humanists' contempt. They are mistaken. [...] Some humanists, on the other hand, are scientifically and mathematically inept and proudly so. Our conversations must turn to matters of their concern, not ours: we must match wits on their turf. As my wife demands: "No physics at the dinner table."

But things are moving, and in many instances they are moving in the right direction. In the core curricula of many universities one may familiarize oneself with chemistry, physics and mathematics in courses like *Physics for Poets*, *At home in the Universe*, *Concepts from Science* or *Science for the millions*. At the University of Amsterdam, together with a team of colleagues from the biology, astronomy and biochemistry departments we present a lecture series called *Turning points in science*, based on the concepts set out in this book, and it reaches a much broader audience than we had expected. In 2000, the member states of the European Union agreed in Lisbon to increase the number of science and engineering students by 15% in 2010. This might be achieved by a number of measures such as stimulating the interest in science and technology of teenagers, and by adapting the curricula, in such a way that these become more appealing to the average student, and in particular to girls.

Other positive signs can be found in bookstores, where popular-science sections are growing, even if they still are rather marginal compared to, say, the section on spirituality, esotericism and alternative medicine. I remember a colleague who couldn't find his own book in the store quipping with a pessimistic sigh: "The Middle Ages are immortal!" On the other hand, the growing number of titles in the science section also bears witness to the growing commitment of scientists to break out of their ivory towers and share their knowledge and opinions with the general public.

The basic problem however arises in high school, perhaps even earlier. What about the elementary-school teacher who had to admit that her adding, subtracting, and multiplication were fine, but that her dividing skills caused some problems? Elementary schools as well as high schools, colleges and universities have a responsibility at all levels to make sure that their students' talents are not lost along the way. This is crucial, because the problem with scientific talents, and in particular mathematical talents, is that they usually get lost irreversibly if they are not recognized and nourished. Few people start learning mathematics when they are 35, while courses in all kinds of management and social skills for that age bracket flourish. There is a strong need for teachers with the real drive to convey the exciting message of science to younger generations. Recall the magnificent English teacher played by Robin Williams in the 1989 movie *Dead Poets Society*, directed by Peter Weir, who succeeds in converting a whole class of impossible adolescent boys into the appreciation of something as venerable as poetry. Not an easy task, I would think, but that is the type of teacher we need for the propagation of science.

I have talked about the mental construct called culture and claimed that there is, in the end, only one science. Both are sublime products of the human mind and together they form another double helix in time, called civilization. Man appears to be a main character in a book of which many chapters still have to be written.

But then arises the doubt – can the mind of man, which has, as I fully believe, been developed from a mind as low as that possessed by the lowest animals be trusted when it draws such grand conclusions?

Charles Darwin

An important counterpart to the message of this book is that we have to live with uncertainty. I think this is a most essential part of living now: at this instant, somewhere along the way, not knowing how far we still have to go or even where we are going. Pessimists may argue that it is like *Waiting for Godot*, the play by Samuel Beckett; waiting for Godot who never shows up … and who was that guy anyway? Optimists may claim that all the answers are already there; one just has to read the great books and interpret them the right way.

The question is how to cope with our *lack* of knowledge and understanding. We live in a house that is still under construction: not only is it unclear whether it will be beautiful, or comfortable, but even whether it will have a roof. To find out we have to go up the stairs but we don't know how many steps there still are to go.

The etching *Donde hay gonas hay mana (where there is a will there is a way)* of Francisco de Goya on the previous page seems to convey a similar message. It reminds us of the great engineer Leonardo da Vinci who worked seriously on designing wings, so that man could lift himself up from the earth surface. On the other hand if we are too eager to fly up too high, our wings may be melted by the sun, and we may fall down, as the Greek myth of Icarus teaches us. We are caught in between; hovering over our own world and existence during day and night, together as well as by ourselves. Slowly coming to grips with it.

It brings to mind the last interview with Richard Feynman for the BBC's *Horizon* in 1981, in which he reflects on his life as a scientist, a touching account:

You see, one thing is, I can live with doubt and uncertainty and not knowing. I think it's much more interesting to live not knowing than to have answers which might be wrong. I have approximate answers and possible beliefs and different degrees of certainty about different things, but I'm not absolutely sure of anything and there are many things I don't know anything about, such as whether it means anything to ask why we're here. […]

I don't have to know an answer. I don't feel frightened by not knowing things, by being lost in a mysterious universe without having any purpose, which is the way it really is as far as I can tell. It doesn't frighten me.

Sander Bais: *The Equations: Icons of Knowledge*

John D. Barrow: *New Theories of Everything*

Robert P. Crease: *The Second Creation: Makers of the Revolution in Twentieth-Century Physics*

Richard Dawkins: *The Selfish Gene*

Michael S. Gazzaniga: *Human: The Science Behind What Makes Us Unique*

James Gleick: *Chaos: Making a New Science*

Brian Greene: *The Elegant Universe: Superstrings, Hidden Dimensions, and the Quest for the Ultimate Theory*

Stephen Hawking: *The Universe in a Nutshell*

Harold J. Morowitz: *The Emergence of Everything: How the World Became Complex*

Simon Conway Morris: *Life's Solution: Inevitable Humans in a Lonely Universe*

Steven H. Strogatz: *Sync: How Order Emerges from Chaos in the Universe, Nature, and Daily Life*

M. Mitchell Waldrop: *Complexity: The Emerging Science at the Edge of Order and Chaos*

James D. Watson: *DNA: The Secret of Life*

List of illustrations